PATHWAYS TO AGILITY

LANCHESTER LIBRARY, Coventry University
Much Park Street, Coventry CV1 2HF Telephone 024 7688 8292

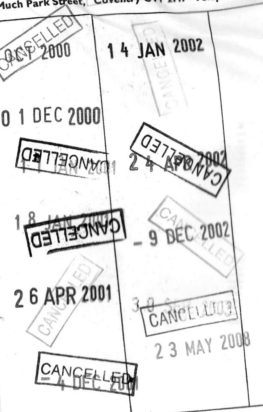

This book is due to be returned not later than the date and time stamped above. Fines are charged on overdue books

PATHWAYS TO AGILITY
Mass Customization in Action

JOHN D. OLESON

John Wiley & Sons, Inc.
New York • Chichester • Weinheim • Brisbane • Singapore • Toronto

This book is printed on acid-free paper. ∞

Copyright © 1998 by John D. Oleson. All rights reserved.
Published by John Wiley & Sons, Inc.

Published simultaneously in Canada.

No part of this publication may be reproduced, stored in a retrieval system or transmitted in any form or by any means, electronic, mechanical, photocopying, recording, scanning or otherwise, except as permitted under Sections 107 or 108 of the 1976 United States Copyright Act, without either the prior written permission of the Publisher, or authorization through payment of the appropriate per-copy fee to the Copyright Clearance Center, 222 Rosewood Drive, Danvers, MA 01923, (508) 750-8400, fax (508) 750-4744. Requests to the Publisher for permission should be addressed to the Permissions Department, John Wiley & Sons, Inc., 605 Third Avenue, New York, NY 10158-0012, (212) 850-6011, fax (212) 850-6008, E-Mail: PERMREQ@WILEY.COM.

ISBN: 0-471-19175-2

Printed in the United States of America.

10 9 8 7 6 5 4 3 2 1

Contents

Preface ix

Acknowledgments xiii

Introduction xv

Part I

Industrial Direction—Agility and Mass Customization

1	**MOVING TO THE INDUSTRIAL SOCIETY**	**3**
	The Historical Perspective	3
	A New Manufacturing System?	9
	Defining Agility's Role	14
	The Importance of Managing Change	16
2	**STRATEGY AND CHANGE IN AN INDUSTRIAL WORLD**	**19**
	Industrialization	19
	Corporate Culture for Agility	20
	Changing with Agility	22
	Necessities for Effective Change	24
	Best Practices—Agile and Custom Manufacturing	32
	An Agile Strategic Planning Process	37

Part 2
Operational Agility and Mass Customization

3	**SUPPLY CHAINS**	**47**
	Today's Business Organization	49
	The Business Supply Chain	51
	Supply Chain Improvement	57
4	**THE EXTENDED SUPPLY CHAIN**	**65**
	Extending the Supply Chain	66
	Customers, Markets, and Channels	69
	Suppliers	77
	Summary of the Extended Supply Chain	80
5	**INTEGRATED INFORMATION SYSTEMS**	**83**
	Manufacturing Process Information	84
	Integrated Product and Process Data	87
	Process Control	89
	Integrated Enterprise	91
	Scientific Computing	95
	Summary of Integrated Information Systems	97
6	**CHANGING CAPABILITY**	**99**
	Today and Tomorrow's Capability	100
	The Customer Order	104
	Customization	105
	Satisfying the Order	107
	The Three Flows	110
	Product and Service Quality	114
	Summary of Changing Capability	116
7	**AUTOMATION**	**118**
	Agile Automation—Process or Machine	121
	Agile Automation—Material Handling	124
	Agile Automation—Information and Decision-Making Processes	125
	Agile Automation—Cash Flow	128
	Discipline in Automated Agility	129

Part 3

Changing Capability in a Modern Enterprise

8	**NEW CAPABILITY—PRODUCTS, MARKETS, AND PROCESSES**	**135**
	The Three Dimensions of Commercialization	136
	Mass Customization—A Part of Commercialization	146
	Summary of Three Dimensions of Commercialization	150
9	**PROCESSES**	**152**
	Process Dimension of Commercialization	153
	Procedure for Change—New Process Technology	158
	New Process Capability—Porsche	163
	Summary of New Process Technology	166
10	**PRODUCTS**	**168**
	Commercialization—New Products	169
	Customized Product—Levi Strauss	170
	Commercialization Process—New Products	173
	Commercialization—Speed to Market	178
	Summary—New Product Capability	180
11	**MARKETS**	**182**
	New Applications	183
	Growth through Regional Market Expansion	188
	Growth through Global Market Expansion	189
	Market in the Commercialization of New Products	191
	Summary of New Market Capability	193

Part 4

Pathways to Agility—The Journey

12	**AGILE RELATIONSHIPS**	**199**
	Cooperation to Compete	200
	Relationships in a Supply Chain	203
	Cooperation between Competitors	204
	The Need for Agile Business Relationships	211

Agile Relationships within an Enterprise		215
People Make the Difference		218
Summary of Relationships		222

13 PATHWAYS TO AGILITY — 224

Establishing the Direction of Change — 225
Pathway to Operational Change — 226
Pathway to New Technology Change—Product, Process, and Market — 236
Summary of Pathways—Operational and New Technology — 242

14 CONCLUSION — 247

References and Suggested Reading — 254

Index — 257

Preface

Pathways to Agility describes how agility has become a part of the next phase of industrial development and how it can be used in manufacturing to move companies toward mass customization. Agility enables companies and individuals to respond effectively to unexpected or rapidly changing events by developing the capability to anticipate the unexpected. Agile companies grow, evolve, and reinvent themselves as they respond to changes in the marketplace and to the needs of their customers. Drawing from my experience with agility in the process industry and as a member of the Agility Forum, I have designed this book to give the reader an understanding of how to apply agility to the supply chain, division, company, operational process, and supplier relationship. Here's an overview of what you can expect to find in this book.

The first section details the history and background of agility and describes the strategic changes necessary to implement agility and mass customization. Chapter 1 looks at past trends, from the hunter-gatherer age to industrialization and ultimately mass customization. The role of agility at each stage is defined and examples of key changes and advancements are highlighted. Chapter 2 addresses the need for strategic change in the industrial world. This chapter discusses why change is necessary in the culture of an organization and the rate of change that can be accommodated under various conditions. When change is not anticipated, the company must make a significant effort to react and respond. However, agile companies have developed the capability to respond to change easily and effectively. To build this capability, the enterprise must have a vision that reflects how it plans to do business, and the enterprise will need progressive managers who understand how agility and mass customization fit into that framework.

The second section focuses on building the organization's ability to respond quickly to change and the operations that support that capability. This section is central to understanding how to apply agility to a company's operations. The supply chain and the extended supply chain must be the focus of the firm's operations. Information systems need to support integrated operations. Chapter 3 focuses on the supply chain, which is viewed as the basic building block within a firm. A business must determine how the supply chain currently operates and what practices must be changed to improve overall effectiveness. Chapter 4 deals with the extended supply chain where the basic business unit is expanded to include customers and supplier. The company must determine what part of the supply chain can be contracted to outside suppliers. Material, information, and cash comprise the essential elements that need to be optimized to make the firm more competitive. Improvement in these elements makes for a more effective operational supply chain. Chapter 5 discusses integrated information systems—one of the key elements of operating an extended supply chain. The use of the computer systems and integrated software can significantly increase the agility of the extended supply chain. The software decreases a supply chain's lead times by allowing information to be entered once and then using it where needed along the extended supply chain. With electronic interchange between the firms along the supply chain, the whole can operate as one entity. Electronic cash flow shortens the cash cycle. The advances in the information system will enable the collection of data from the shop floor or the process, which can then be used to improve the manufacturing operation. This in turn allows for the development of new process and product technology, as well as enabling the personal service to customers and suppliers that leads to mass customization. Chapter 6 deals with the need for the capability to enable improvement. Several examples show how changes in the structure of the supply chain allow for an agile response to unexpected events. Chapter 7 discusses how automation can support an agile supply chain. The automation of the movement of material has a significant impact on how agile a firm can be. Automated information and decision processes, as well as machines, processes, and material movement capability, must be designed to anticipate unexpected change. Automation promotes discipline in the operation of the supply chain, which in turn enables agility by allowing error-free production, even under unexpected circumstances.

The third section of the book looks at changing capability in the enterprise relative to the needs of new processes, products, and markets. These elements are the company's lifeblood and any changes to them must be carefully considered. Chapter 8 begins by looking at change that involves all three elements. The Dimensions of the Commercialization cube is introduced to illustrate the complexity of the effort. Chapter 9 describes the integration of new processes into a company's operations. The mitigation of risk is discussed relative to the need for change. The company must have the right manufacturing capability to enable an agile response and at the same time lower risks significantly. Chapter 10 focuses on the specific requirements needed to introduce new products within the commercialization model. The company must develop techniques to prevent the introduction of the new product from overcommitting the firm. Chapter 11 looks at market expansion where effective actions at the supply chain and cross-functional team level can ensure success and minimize risk. The commercial venture that simultaneously introduces a new process, new product, and new market must take special care in choosing the right approach in order to succeed in this high-risk scenario.

The fourth section of the book deals with pathways to agility, including the need for agile relationships with suppliers, fellow team members, and even your competition. Chapter 12 explores the relationships that a firm must have to function effectively and the importance of people both in and out of the organization. People make the difference in a company's operations and getting them to work together is essential to success. Likewise, agile relationships with your suppliers and your competition can lead to increased effectiveness within your own company. Chapter 13 lays out the pathways to agility and the change process. After applying agility within the organization, you will find that the pathways merge and change and become second nature in the firm. Chapter 14 concludes the book and ties together key points.

Acknowledgments

The support and encouragement of my dear wife, Carol, helped immensely to keep the positive approach to writing this book. She helped provide the atmosphere that allowed for capturing the unique and creative aspect of the book. My father, Douglas, who is retired and living in Florida, is to be thanked for providing a quiet and comfortable atmosphere for intensive writing.

Colleagues Bob Chapman and Jack Boone labored hard to get through the initial writing and helped turn it into an understandable text. Jack Allen provided input when he could that was both creative and stimulating.

Dow Corning provided encouragement and support to help make the book a reality. The Agility Forum, particularly its members Rick Dove and Sue Hartman, also provided the opportunity, through visits to companies and discussion, to gain an understanding of the concepts of agility and mass customization.

Introduction

Change has occurred in commerce since the beginning of time. In modern times we've focused on new manufacturing methods, shifting from mass to lean production, and are now at the next wave of manufacturing innovations: *mass customization*. The manufacturing industry is taking on a new shape and that shape is significantly impacting how business is done, emphasizing the activities of the whole enterprise and relationships across the extended supply chain, including both the suppliers and the customers. New relationships with customers and suppliers are changing the nature of the entire supply chain. These relationships may take the shape of a partnership to develop a special type of new product. They may involve suppliers being asked to improve their economics and to pass the savings on to the customer. They may focus on the relationships that existed between the "silos" of activity along a supply chain within an enterprise. The change may be outsourcing—the contracting of activity that a firm elects not to do, turning it over to someone more skilled in the activity.

One key element within all of this activity is the practice of *agility*. Indeed, agility is necessary for the successful implementation of these new relationships and commercial activities. At its most basic level, agility is the ability to respond effectively to unexpected or rapidly changing events. As the nature of business changes, firms must respond effectively. On one hand, the response might be reactive. More recently, however, the response is seen as a *capability*. Business must shift toward planning for unexpected change by anticipating it. This approach is the theme of this book—expecting and preparing for the unexpected.

Whether you are a senior manager, a production manager, or a line supervisor, you may be looking for a simple answer to the question "What is agility?" When something applies as broadly as agility, however, it is difficult to define it out of context. It is more useful to talk about agility in specific circumstances. But for the purposes of this book, agility is defined as *the ability to respond with ease to unexpected but anticipated events*. That is, the capability has been established that allows for a response to be executed with ease.

I first encountered agility at a 1992 American Society of Mechanical Engineers (ASME) show in Detroit where a special program on the topic was presented. About 20 people were in attendance. The other attendees were from the auto industry; I was the sole person from the process industry, coming from a chemical company. From the outset I could see that agility applied both to the discrete and process industries, but I wanted to know more and immediately became involved with a team that assessed agility at various industrial companies. In the beginning, the assessors were learning as much as the companies being assessed. However, progress was rapid and each visit gave us additional insight into how agility could be applied. The teams were led by Rick Dove of the Agility Forum.

From observing Rick on these visits, it was apparent that he spent a lot of time learning from others and then was able to translate the information he had gathered into concepts. With his leadership, the traveling team put together an agility assessment criterion that has been very valuable. The value came from both the process of developing it and the criterion itself. However, I was still left with the need to translate agility for the process industry. The task was more difficult than it initially appeared. Agility did not seem to completely fit the context of either the discrete or process industry, although it was being promoted by government and industry leaders as the next wave for industry after lean manufacturing. It was only later that I realized that agility has been required in all the waves of industrialization—from the crafts worker stage to mass production to lean manufacturing. And as industry moves to mass customization, agility will become even more crucial for the success of an enterprise. Anticipating the unexpected from the customer and then having the capability to custom manufacture a product will take real creativity. Mass customization will not happen unless the customers' needs are anticipated and the capability to fill those needs is developed.

This early work enabled me to see how adapting agility for the process industry was possible. In my company, I was involved in developing a vision for taking our manufacturing establishment into the future. Our vision for large continuous processes was oriented toward lean manufacturing and economy of scale. However, I eventually realized that even large-scale continuous chemical processes require agility. These processes required that a mix of a half-dozen products and by-products be balanced to meet the level of demand from the customers. These demands came from further down the supply chain in the finishing sector of our business. Over the years, while establishing lean, low-cost manufacturing operations, the technologists had built the capability to balance the by-product production. When that capability was exceeded, the low-profit portion of the product line was phased out and the appropriate shift to higher-profit products occurred. As a result, the finishing part of the business became agile and our vision for the enterprise emphasized the ability to respond to the unexpected demands of our customers in an agile fashion.

This initial vision set the stage for further development of the applications of agility. By focusing on agility it was possible to develop mass customization for many of the product lines and supply chains. Lean production allowed the streamlining of supply chains, and mass customization provided the ability to serve the market as customers demanded. Agility provided the necessary framework to bring the two together. This change in direction to lean production and mass customization has enabled the corporation to become much more competitive. What started as a conceptual exercise has resulted in our vision being printed in the *Green Book*, which lays out the overall direction for the manufacturing divisions and supply chains of Dow Corning. This book, which defines relationships between entities within the corporation along with the measures and metrics of key goals, sets the direction for change. Supply chains, sites, and departments tailored it to their needs and used it to provide overall direction, quantifiable goals, metrics, and measurements. It served as the guiding force as the company's focus shifted to supply chains and lean and custom manufacturing. It provided a framework for learning and applying new concepts across the organization.

These similar experiences from a wide variety of companies provide the basis for *Pathways to Agility*. The book is intended to act as a bridge from concept to practical application of agility in the industrial world. It will describe the pathway that an industrial firm should follow to effectively implement agility into its activities. It will also highlight the benefits that agility brings to a company when integrated with existing and new commercial activities of the firm. As you read this book, you should interpret agility concepts in the context of your own circumstances whether they be in the discrete or process industry. They apply to both. Remember that while the principles will apply in most situations, change is not easy to initiate or implement. Keep in mind that agility is a tool that must be included in the improvement activity of your firm, especially as your company shifts from mass or lean manufacturing to mass customization.

With this background in mind, I will briefly look at the history of agility and how it has developed into the current model. The changes that have gone before form the foundation for today's concept of agility and influence the impact that agility will have on the future of manufacturing.

PATHWAYS TO AGILITY

PART I
Industrial Direction—Agility and Mass Customization

CHAPTER 1
Moving to the Industrial Society

The Historical Perspective

The history of agility goes back to when hunter-gatherers were pursuing food. They needed to outsmart their prey and react to unexpected events to survive. Their days were filled with unforeseen events as they encountered the hazards and opportunities of the day. It was not a controlled environment, but instead one that required the hunter-gatherers to build the capability that allowed them to meet the unexpected. The more agile were successful in this pursuit.

Agility continued with the farmer who needed to be prepared for and respond to unexpected changes in the weather, to learn the practice of effective agriculture. In the beginning, farmers did not know the best time to plant and harvest, but eventually learned by responding to changing conditions. Uncertainty in agriculture, where unpredictable markets, weather, and crop yields affect the outcome, caused the farmer to become agile in his approach. In most cases farmers have been successful in adapting to uncertainty and producing adequate harvests. Now, the uncertainty has shifted from the impact of Mother Nature to variations in the market.

A crafts worker was a custom producer who made a single unit for a single customer tailored to the specifications developed for that customer. This practice applied to everything from furniture to clothing. The town blacksmith worked with forge and metal. The furniture maker worked with wood. The tailor worked with cloth. Specialization went further with the brewer making beer and the seamstress making women's clothing while the tailor made clothing for men. Each individual craftsperson had a specialty but adapted the product to the requirements of the customer. This did not mean that a wide variety of products was available but the adaptation occurred within the capability of the individual crafts worker. Agility and the ability of crafts workers to accommodate the needs of their customers have been characteristics that have kept craft important in our lives.

Society has seen the transition from the hunter-gatherers to the agricultural age and the dominance of the crafts worker to the industrial revolution where manufacturing became a prime driver. The industrial age moved to a *mass production society*—where workers were not providing the skills of their hands as much as they were providing the skills of their heads. Workers determined how the machine would do the job. They were also working in a disciplined, assembly-line world where many of the tasks were routine and did not require the skill of crafts workers, who had used tools, mind, and hand to produce the product or service. Industrialization was a continuously improving concept. History has recorded many surges in the rate of change or improvement.

Eli Whitney's invention of interchangeable parts was a major milestone in the evolution of the crafts worker to the concepts of mass production and worker specialization. The military people did not realize that the boxes of parts in the Whitney shop were practically rifles; they expected to get a few handmade and assembled rifles every week. This caused frustration on the part of the military procurement people. Assembling muskets from parts that were made to the same specifications started a concept that became the foundation of mass production.

Mass Production

Mass production accelerated the industrial revolution and put an automobile in everyone's garage. Henry Ford perfected the concept of interchangeable parts starting in the early 1900s. He added the moving assembly line, with workers who focused on tasks defined for their workstation. The worker no longer needed elaborate training for the task, and the division of labor that was characteristic of the craft era was taken to the lowest possible level. A worker would perform a focused set of tasks that did not require broad or extensive training. Rather, this move to mass production required a skilled, knowledge-based worker who perfected the way things were done. These knowledge workers not only perfected the individual tasks at a workstation but also integrated all the tasks into supply chains and assembly lines. Thus the foundation of mass production was established and became the technology that lasted for more than a half a century. It was the king of the competitive world. It ruled the manufacturing enterprise and became deeply ingrained.

Mass production improved significantly after the early 1900s and became the method that resulted in the highest productivity in the industrial world. It was the method of choice for high-volume manufacturing. After World War II, however, for some countries, like Japan, high volume was not possible. Japan did not have a mass production economy as it struggled to establish itself in the industrial world. A different approach was needed, and a revolutionary concept grew out of necessity. The Toyota Motor Company, which originally was named Toyoda after the founding family, started in 1937. It produced trucks as directed by the military government. The passenger car was a dream, but it had to wait. The trucks were produced using the craft method. After the war, the company struggled to establish itself as a force in vehicle production but had produced less than 3,000 passenger cars by 1950. Its struggle appeared to be unsuccessful, and the company seemed to be on its way to extinction. It had worked for 13 years to build those few automobiles and some trucks and was floundering. Something had to be done, and Eiji Toyoda set off to study the workings of the Rouge Plant of Ford. He concluded, however, that the Ford system would not work back home at Toyota; he needed something that worked for smaller-volume production.

Under the direction of Eiji Toyoda and Taiichi Ohno, the Toyota Production System was developed. It focused on quality, productivity, rapid changeover, multitask machines, and economics. It became the envy of all those in the auto industry and beyond, beginning the revolution that changed the approach to manufacturing. It became know as lean production.

LEAN PRODUCTION

Lean production challenged many of the techniques and paradigms that were deeply rooted in the modern mass production society. As with any change, the concepts were not easily adopted by others, especially in the Western world. The global manufacturing establishment needed to learn the principles of lean manufacturing and then needed to adapt them to their situation. People needed a significant amount of time to understand just what made up all of the important elements of this system. It was difficult to focus on product design in a world that was dictated by ease of production, and it was difficult to integrate quality concepts into all of the firm's activities. Integration of function was not readily understood in cultures where the individual was paramount. Japanese culture focused on people contributing to the group with an authority figure in the lead. Being part of the group was of great importance to the Japanese individual. Hard work and competing with the world for the good of the country was also a dominant characteristic of the Japanese workforce of the time. Change occurred because of these motivational forces and was not handicapped by the needs of the individual. Lean production emerged from that world as a force to be reckoned with.

The Japanese presented some different but interesting situations that would not be understood by the Western world. For example, while I was working to build a plant in the Chiba Prefecture, east of Tokyo, a Japanese engineer had a "better method" to provide heat and energy to a process unit. He convinced the group of people that the new design had significant benefit and that it would work. Design and construction proceeded and start-up was under way. However, the new heating system did not work as planned, and the Japanese engineer was under stress. An interesting thing happened. All the members of the group left the job of designing and managing the construction of the plant and joined the start-up team. The group persisted until the new design worked. At the next meeting the Japanese engineer bowed humbly and said *domo arrigato* (thank you). It was not clear what the consequence of the design not working was, but it was probably much more drastic than what would have occurred in the West.

Lean production became the dominant method for Japanese manufacturing. Although the culture that drove lean production made it very difficult to adapt the techniques to companies outside of Japan, the adaptation process did begin in the 1970s with the auto industry. Ford Motor Company had programs of integration aimed at concurrent engineering. Quality is Job #1, and design for manufacturing. Time for new model introduction was being reduced from five to six years to three or less. Product quality was improving, and initial defects were reduced. Workforce productivity increased as designs for manufacturing techniques were introduced. The supply chains shifted toward just-in-time (JIT) delivery to assembly plants, and suppliers were more closely integrated with the company. They were asked or were directed to improve the efficiency of their operations and deliver components at a lower cost with a higher quality. Companies moved from doubting whether the Toyota System would work to learning and adapting its concepts and principles in order to compete. Thus, the lean production concept had become a part of the industrial world in the West. It built on the concepts of mass production and used the technology developed in Japan to change the way things are done.

The movement from the hunter-gatherer to crafts worker to mass production has been greatly simplified in the preceding discourse. Likewise, the advance of lean production has been condensed. It is not the purpose of this book to develop these concepts in detail. However, it should be recognized that each of the advancements had many people contributing, and people like Eli Whitney, Henry Ford, and Eiji Toyoda all used the thinking of others and adapted that thinking to their enterprise. Their contributions were significant, and they represent the technology change leaders in the era in which they excelled. The same will be true in future advances of manufacturing technology.

The changes mentioned in Figure 1.1 focus on manufacturing *technology*. The changes that have been occurring in more recent times stress the efficiency of the *whole enterprise*. Automation of both the manufacturing process and material movement has been occurring throughout the development of modern manufacturing technology.

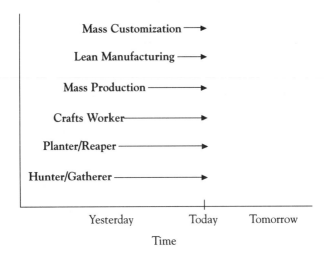

Figure 1.1 Stages of Development

Concurrent activity was demonstrated in the assembly lines at Ford in the 1980s, where parts were being produced at the same time autos were being assembled. This concurrent activity had not reached the knowledge-based parts of the firm. Design was sequential with one step being completed before another. The ability to rapidly introduce new models by the Japanese automakers alerted Detroit that the linear approach needed to change. Ford's program of concurrent engineering resulted from this. To accomplish this, the Ford people focused on the workflow processes and how they fit together. The workflows dealt with the design and development of new Ford models. They needed to shift from sequential engineering to parallel, or concurrent, engineering. Ford found that many tasks could be performed during the design that did not have to interface with others and thus could go on while the critical elements were getting integrated. The parallel approach to the design of a car was based on the agility designers developed as they worked on their part of the car. This meant that it was necessary to anticipate what others were doing. A high degree of agility was required to adjust the designs so that the parts fit and nested together. The ability to anticipate what others were designing was required to make the system efficient. The result was a significant improvement in the time to market for a new model.

Lean production used many of the principles of mass production but made a significant improvement that led to Womack, Jones, and Roos coining the phrase *lean production* in *The Machine That Changed the World*. Lean production became and continues to be the wave of industrial improvement, and firms are adapting to lean concepts. In Womach and Jones's next book, *Lean Thinking*, they extended the lean concept from focus on production to focus on the enterprise. However, they could not answer the question of what would come after lean. They saw, as they toured the world promoting their *Machine* book, that so much needed to be done to get lean principles adopted, that they could not envision the next wave of change. This is where agility and mass customization enter the scene. They are not just an extension of progress but impact all parts of the industrial activity. Mass customization is an extension of the work of the crafts workers. To influence mass production and lean manufacturing as people add capability to offer more variety or quicker response to the changing needs, agility is required.

A New Manufacturing System?

The competitiveness of U.S. industry has improved significantly over the last decade. Much of this has been due to the changes in information systems and the reengineering of workflows. Continued progress on the efficiency of the manufacturing processes has also contributed to U.S. industrial competitiveness. Another force has been the U.S. government's recognition that it can have a significant impact on how effective industry can be. Agility has been a important part of the improvement and will continue to be an important factor.

Although agility has existed throughout the ages in the various steps that define the progress of humankind to the present industrial state, its existence occurred to various degrees out of the necessity of the situation and it did not stress *preparation* for the unexpected. That is, agile responses to a situation occurred, but they tended to be coincidental. As mass customization emerged in the late 1980s agility became a more prominent part of industrial activity. It was detected in both lean and mass production as competitive pressures demanded more variety and a better fit to all customer needs. Mass customization continues to grow as industry develops the concepts of agility.

The U.S. Congress had been worried about the U.S. industrial competitiveness for many years. The surge in competitiveness from Japan with lean production had taught U.S. industry that it was not as good as it needed to be to compete. The 1984 Cooperative Research Act, which allowed companies to cooperate noncompetitively on the development of technology, was a response to this concern. Congress also provided the seed funding to catalyze the cooperation between industrial enterprises. One of the best examples was the formation of Sematech in 1987, which brought the U.S. semiconductor industry together to develop the tools of production. The electronic industry was rapidly moving out of the United States, and companies were not committed to developing the tools needed to keep it in the United States. At first the electronic companies were reluctant to cooperate. However, it quickly became evident that this approach was going to have an impact and enthusiastic participation increased. This stimulated a resurgence of confidence that made the U.S.-based electronic industry surge at the expense of foreign competition.

Indeed, the United States became a place of choice for foreign companies in the electronic supply chain. This project succeeded because of cooperation among industrial companies that initially were reluctant to work together. As a result, industry realized that it needed to improve to succeed, and cooperation was possible to leverage the technology development activity. The ability to work together for the good of U.S. industry led to some other examples of cooperation that are emerging as contributors to U.S. competitiveness. Congress has authorized putting seed money into programs in textiles and automobiles. These programs are called AMTEX and USCAR. The normal inventory investment in the textile supply chain had been very high with lead times of over a year. The thrust of AMTEX has been to focus on reducing textile companies' lead time and to produce textile goods more effectively with a closer integration with the customer. It was also aimed at keeping the whole process and supply chain in the United States.

An example of what this industry faced prior to AMTEX involved Du Pont, a material supplier to the textile industry. In the early 1980s Du Pont found it could not compete with Toray Industry in the markets of Asia with its fibers. It devised a strategy—integrating the supply chain from fiber to garment using computer information systems—to bring the manufacture of textiles back to the United States where the competition would be more balanced. By knowing what was being sold, Du Pont and the U.S. textile industry could produce the fabric and the garments much more quickly than their Asian counterparts. The improvement of the operation of the supply chain brought certain types of clothing manufacture back to the United States.

Case studies were written on the agility that was possible in order to keep up with an industry that changed from one fashion to another. These changes were made much more quickly and the Asian source could not respond. The shortening of the supply chain by providing knowledge of what customers were buying made these companies much more competitive. It was effective enough that clothing produced in Asia was shipped to the United States by air by garment producers. The time to market was that important.

With Du Pont's evidence of the value of agility and a responsive supply chain, the U.S. government funded a project aimed at making the U.S. textile and clothing industry as a whole more competitive. This effort involved coordinating information, providing new methods and tools of production, and operating the overall supply chain in a more integrated fashion. The textile industry was aiming its success at having production occur in the United States, where the pay may be $10 per hour, and compete effectively with other economies where the wages are much lower. This effort was titled AMTEX and continues today.

As mentioned earlier, the AMTEX competitive improvement thrust is to reduce lead times and to produce more effectively; AMTEX strives to accomplish this by applying advanced technology to textile industry problems. This work will provide the machines and controls that change the nature of the industry. Technological innovation is being applied to the textile industry. These are long-term programs that are funded by both industry and government with the goal of a more productive and competitive U.S. industrial society. Complete success has not occurred yet in this program, but some very interesting advances have been made and are beginning to be implemented.

Another example of bringing companies together with government seed money is in the automobile industry. The USCAR program was created to share the development of new technology for the industry as well as to improve its product and process technologies. Industry and government leaders felt that more progress on things like fuel cells would yield results faster if the technology was shared across the industry as opposed to being developed secretly in each organization. If success occurs, the competition would be in the marketplace as each firm takes the technology into commerce. This is being done at the big three automakers and their suppliers. The first hints of success are beginning to appear. The hope would be that fuel efficiency can be greatly improved and that transportation can take a lesser share of the global natural resources. Improvement through teamwork can indeed make the U.S. automobile industry more globally competitive.

These programs of cooperation used the catalyst of government support and finances to bring some reluctant industry participants together for the sake of industrial competitiveness. Traditionally, it has been a part of U.S. culture that competition was good, and if competitors got together and shared anything, it would be looked upon as a restraint of trade. The shift to a more global economy, however, made that philosophy obsolete. There was significant competition from overseas. Our governmental leaders wisely perceived the situation and legislated a change. They not only built an enabler into the law for cooperation but also provided initial funding to break down the resistance or timidity of companies to cooperate. The 1984 Cooperative Research and Production Act will probably go down in history as a major piece of legislation. It is a success story, and industrial cooperation is the driver that will make the competitive improvement possible. Agility was a part of this drive to make the nation more competitive. The atmosphere that agility emerged in was one of intercompany cooperation and change or improvement in the way industry worked. Technology in the production processes and supply chains was the focus, and government was providing the leadership for the change.

In the spirit of enhanced U.S. industrial competitiveness that had emerged in the late 1980s, the Secretary of Defense Manufacturing

Technology (Mantech), through the Office of the Assistant Secretary of the Navy, searched out a way to redirect industry toward agility. They found the seeds of the current concept at Lehigh University, which had initiated work on industrial competitiveness. The government sponsored an effort in 1990–91 to better define what the next enterprise strategies should be. Dr. Roger Nagel of Lehigh University's Iacocca Institute was contacted and the initial effort to define the strategies was started. The effort was led by Roger Nagel and Rick Dove as coprincipal investigators. They brought together a team of industry executives from a broad segment of the U.S. industrial world. Within a short time, the topic was brainstormed and creatively developed. It resulted in a two-volume publication titled *21st Century Manufacturing Enterprise Strategy*. Editors of the two volumes were Steven Goldman and Kenneth Preiss, both from Lehigh. This effort resulted in the coining of the title of what the next industrial change would be. They used the term *Agile* to describe the key characteristic of competitiveness. The key points that were summarized on the first page of volume 1 are shown in Figure 1.2.

These key points highlighted the changing demands of the consumer and others along the supply chain. Expectations were being developed that required a different response from the service or product supplier. The enterprises that responded would have the competitive edge. Higher quality and customized product and services with competitive prices would increasingly become the norm, and enterprises would need to change to meet the expectation. These key points are as valid today as when they were developed in the volumes that coinvestigators Dove and Nagel put together in November of 1991.

They also called attention to a role for the Department of Defense (DOD) as an agent of government to play in making the transition to industrial agility. In 1992 the National Science Foundation (NSF) and the Advanced Research Project Agency funded a proposal defining the initial structure of the Agile Manufacturing Enterprise Forum, which is discussed in the next section. It was an organization driven by a passion for enhancing U.S. industry's competitiveness. The consequence of not advancing to new manufacturing and service technology and competitiveness would be to put the U.S. standard of living at risk.

> **Figure 1.2 Characteristics of Competitiveness—Agility**
>
> - A new competitive environment for industrial products and services is emerging and is forcing a change in manufacturing.
> - Competitive advantage in the new system will belong to agile manufacturing enterprises, capable of responding rapidly to demand for high quality, highly customized products.
> - Agility requires integrating flexible technologies of production with the skill base of a knowledgeable work force, and with flexible manufacturing structure that stimulates cooperative initiatives within and between firms.
> - The Department of Defense has a vital role to play and has a large stake in the success of industry in making the transition to agile manufacturing.
> - The standard of living Americans enjoy today is at risk unless a coordinated effort is made to enable US industry to lead the transition to the new manufacturing system.
>
> Source: Steven L. Goldman and Kenneth Preiss, eds. *21st Century Manufacturing Enterprise Strategy: An Industry-Led View.* Bethlehem, PA: Iacocca Institute at Lehigh University, 1991.

Defining Agility's Role

Agility was being put forth as a new manufacturing system that would be the next wave, or stage, of industrialization. However, unlike former industrial advances, agility can exist within the mass production or lean stages of industrialization. This intellectual conflict has made it difficult to understand what agility really is and how to manage it. The important lesson that I have learned through years of involvement is that agility exists in all stages of industrialization—crafts worker, mass, lean, and now custom manufacturing. A more agile industrial world will emerge with an increasing competitiveness that occurs because of many improvements across all of industry and society, one of the more important ones being the creation of an industrial workplace that can easily respond to unexpected but anticipated change. The Agility Forum is playing a key role in defining how agility can contribute to the competitiveness of industry. Some of the Forum's history and present activities are presented in the next paragraphs.

The activities of 1990 through 1992 led to the establishment of the Agility Forum. The Agility Forum was initially funded in the fall of 1992 by the National Science Foundation and Advanced Research Project Agency. The idea and concept were deemed creditable and further participation from industry resulted. Texas Instrument became an early participant, although involvement quickly expanded with representation from companies across U.S. industry. A research-oriented structure evolved and much discussion, interaction, and many site visits have occurred. Each participant strived to obtain a better understanding of agility and how it was or was not working in industry.

In the early years of the Forum's existence, I became involved in the activities and made many trips to see what others were doing to make their enterprises more agile. It was here where I saw the need for doing business differently from the way it was being done. In each visit I gained new knowledge; I saw my role as adapting that knowledge to my firm in the process industry. Benchmarking was critical to understanding the practical side of agility. This early sharing among companies was a very important role of the activity. The visitors not only saw what the host was doing but also talked among themselves and built an understanding of direction and difficulties that each was experiencing. The whole effort allowed for the development of a database of case studies that are extremely valuable in managing change. Many of those experiences are captured in this book and reflect the personal experience of the companies we visited.

The Agility Forum provided a focal point for many of the efforts that are working to make the U.S. industrial enterprise more agile. Indeed, agility is creeping into the capabilities of many firms and the case studies of good practices are expanding. As in the other activities from hunter-gatherer to lean manufacturing it becomes a blend of activities until the proper emphasis is arrived at. The agility effort, which came out of Congress's desire to increase U.S. industrial competitiveness, has established a firm foundation and is spreading, but it needs further emphasis and must become a key change tool or attribute at most corporations with a focus on the supply chains that they run. Much needs to be developed and written on the successes of the concept, because it may still be classified as a phenomenon that is searching for understanding. Agility exists in all the elements of industrialization. It has existed with the crafts worker, mass production, and lean manufacturing, and it will be required in mass customization. Agility will continue to exist across industry. The discussion of agility needs to be expanded so that companies understand how agility gives an enterprise or a supply chain a competitive edge. Writing the history of agility is not finished. It will be continuously recorded as U.S. industry enhances its ability to compete. Agility will be a key element in the continued drive to mass customization.

The Importance of Managing Change

Change is a crucial factor in implementing agility and mass customization. The changes that the business world will encounter will be very significant. Firms must shift from a reactive to a proactive mode to set the stage for how business will be done within the segments where they compete. Firms cannot sit back and watch something happen and then react to it. They must be looking at other industrial segments to see what is working and whether they should adopt a particular business practice. The concept of benchmarking outside one's area of business will become increasingly important as companies shift to become more agile operations. Since all change builds on the improvements in the past, it is important to understand their significance and apply modern thinking to changes of today. A significant challenge will be to see how the information age will influence the future. Concepts of automation of information flow will require foresight. Other trends will need to be assessed carefully to see how they apply to firms for the future.

An important lesson is to ensure that one is not stuck in a paradigm that looks back. The change agent or business strategist must be forward thinking. Ian Morrison and Greg Schmidt in their book *Future Tense* described that frame of mind in the box that follows.

This concept of using foresight and applying it to the business situation and strategies of a firm must be kept in mind. Organizations need people who are setting the direction and making the big changes that will make the difference. The continued improvement of competitiveness is essential to the future of the firm. Agility and mass customization will be essential to the future competitiveness and combined with other trends should provide a change strategy that has the firm moving to "where the puck will be." The Gretzky Effect (Figure 1.3) should be remembered and used as part of the direction setting for change; that is,

---

Foresight and the Gretzky Effect

"Wayne Gretzky is the greatest ice hockey player of all time. Not particularly big, strong or fast, Gretzky is arguably the most successful athlete in any sport. When probed on the reasons for his success, he responds: 'I skate to where the puck is going to be.'

"Gretzky has foresight—an uncanny ability to see a complex reality unfolding in front of him. He has a video player in his head that gives him a fast forward look at the future. Such a gift is rare in athletics. But we would argue that in business the same ability can be acquired and learned. Foresight, the ability to anticipate how a complex world will unfold, is becoming an increasingly important skill for business people. We are not talking about predicting the future. That's impossible. But thinking systematically about how events and trends unfold is not only possible, but necessary. When applied by someone with good business instincts, knowledge and experience in industry, and an ability to analyze and forecast, the results can be remarkable."

Ian Morrison and Greg Schmidt, *Future Tense: The Business Realities of the Next Ten Years*. New York: William Morrow, 1994.

"I skate to where the puck is going to be."

Figure 1.3 Foresight and the Gretzky Effect

change programs must anticipate the events and timing of the future to lead the firm to a more competitive position. However, foresight must also be looked at as being well ahead when the change is complete but not so far ahead that others, like customers or suppliers, are not also a part of the change.

This chapter has introduced the importance of preparing for the strategic change that must occur within a firm to make it competitive in the future. The next chapter will talk about the strategy of change in an industrial world as well as the corporate culture and its ability to change. The chapter will focus on the processes for change management. It will deal with assessing the best business practices and determining how a firm desires to do business. It also will discuss agile and custom manufacturing as one of those business practices. Last, the chapter will describe an agile strategic business planning process that includes the strategic change elements as part of the operational activity of the firm.

CHAPTER 2
Strategy and Change in an Industrial World

Industrialization

The industrial world is in constant change. To the outsider it appears chaotic, but it is usually well orchestrated and managed. The people proactively involved in change in a corporation have a self-prioritization system that looks at all the opportunities for change or improvement and determines which will get the most results, the fastest, with the least effort, and they rise to the top of the change priority. Those opportunities that do not have apparent benefit are usually mentioned but little effort or results occur. This is regardless of the source of the change idea or thrust, which can come from the top down or be initiated by people with good ideas. All change will be treated the same in the long run. It is this world that the concept of agility and mass customization faces when it becomes a change thrust. The industrial world has asked, "Who needs it?" The response to the question has not been easy. The answer usually requires a description that takes one into the distant future. It also requires an integrated change program across a domain that individuals being asked to be part of the change cannot see. It raises the question, "If I don't see others changing, then why should I?" This presents the advocates of change to a more agile or mass customization corporation with a significant problem or challenge: how to get significant change that spans the corporation or a supply chain.

In calculus, the equation defines a place where an object resides. The first derivative is how fast the object moves. The second derivative is the acceleration of the object as speed is increased. Agility as a concept is like the second derivative. Change occurs in the corporation. Agility is the acceleration of that change. Competitiveness improvement is the primary driver of change within a supply chain. This book provides the reader with an understanding of the language for those issues that are changing manufacturing and discusses how they apply to making an enterprise more agile and thus more competitive.

Corporate Culture for Agility

The concept of change in a corporate culture varies with the history of the firm. An example of a company that has a culture that accepts change is Motorola, noted for its ability to rapidly respond to unexpected change. One of Motorola's most profound change mandates was six sigma quality—now the tenet of success for everything that Motorola does. Motorola instituted six sigma quality ahead of most industrial firms. At the time this change was implemented, the company was operating "under the shadow of the gallows," that is, the company was having trouble competing and was downsizing personnel at all levels within the company. Initially, Motorola employees felt instituting six sigma quality could not be done and that the company could not afford the cost of this level of quality. The leadership, which had done benchmarking of the competition and other firms in a variety of industries, insisted that it would be done because the future of the company depended on it. Thus the change was implemented and six sigma quality became an operating paradigm at Motorola. In addition, management saw how effective implementing the change was while the company was in a crisis atmosphere, and the concept of being "under the shadow of the gallows" became a part how Motorola manages change for all its projects. In this example, a corporate culture shifted to a paradigm in which being the best in whatever the company did became paramount. Rapid change and six sigma quality are now embedded in Motorola's culture and are the foundation of this dynamic organization.

ENTER AGILE MANUFACTURING

When the U.S. defense industry was reducing its size in 1992 as the cold war came to an end, Motorola was challenged to change the way it supplied the electronic boxes that went into various weapon systems. The defense establishment had insisted on dedicated manufacturing lines to produce electronic system electronics. The electronic systems were composed of specially designed electronic boards that were integrated into a system that was the brains of the various weapon systems. To produce these systems Motorola employed a lot of people on the various lines. Each line specialized in a weapon system. With the end of the cold war, weapon purchases by the government declined. Faced with the prospect of a significant reduction of orders, Motorola leaders set out to change the way they respond to this business. They suggested that the Defense Department had two choices. It could buy all the new systems it would need and place them into inventory, or it could support the development of an *agile manufacturing* system. Both Motorola and the government agreed that the best choice was an agile manufacturing capability. They set out to establish this capability in an atmosphere of creativity and urgency. The project was conducted "under the shadow of the gallows," as the shrinking business would dictate. They shrank the size of the operation and changed it so they could effectively make just a few electronics at a time as the orders dictated. This change involved automated agility from material logistics to component placement on boards and machine instructions, which significantly changed the way this part of the defense weapon industry did business. Motorola thus had responded with a very effective new agile capability designed for custom manufacture. The manufacturing capability came on line in six months and was operating at six sigma shortly. Physical size of the facility was reduced significantly. Lead time in the supply chain was improved with a make-to-order and a just-in-time philosophy. Empty buildings and idle people were shifted to cellular phone production to take advantage of these buildings and the people's skills. The results have been a real success story for a very successful company. Agility was brought to a highly technical application in a controlled but rapid fashion.

Changing with Agility

The preceding example points out some key elements that drive change within corporations. In the case of Motorola, the culture responded to a significant challenge that was affecting the livelihood of the enterprise. The enterprise had felt the pain of not being competitive and had seen the success associated with six sigma quality. The culture was make-it-happen oriented and was engraved in the company with significant training for all employees. Motorola is a corporation where employee training is a paramount part of the culture. Motorola University exists around the world and teaches the necessities of keeping the company successful.

Not all corporations have change as a part of their culture. The management of change is a very important capability for any corporation. The amount of change that the leadership of a corporation desires must be balanced with the amount the organization is capable of handling. If the desired change exceeds the capacity for change, then the organization will prioritize. This will result in certain changes being randomly dropped. An example of this occurred across industry a number of years ago. Industry was busy adopting the concept of total quality. Slogans were developed, and programs to improve the quality of everything that is done or produced were a significant trend. This went on for several years, but the results were disappointing. The overall success that was hoped for was not achieved because industry did not add the resources necessary to conduct the programs to improve quality. Many companies saw that the measurements of success could be achieved by selecting some "low hanging fruit" and making selective improvements. The exception to this is where the change was essential to being competitive, as in, for example, the automobile and electronics industries. Other industries did not feel the same competitive pressures and gave the effort to become more competitive a lower priority. Thus, the significant change that was desired did not occur.

Change is prioritized randomly at various places in a corporation when individuals determine what is important based on the activities and situation they are experiencing and then decide what will be done. Implementing change randomly, however, can have a very negative effect. For example, if a particular change is needed at all sites in the supply chain, random implementation can cause time delay and/or failure of the change. People will wonder what happened. Why didn't the change occur across the supply chain or enterprise? Why did the change leadership think it was happening but in reality it was not? Who changed the priorities? Where did the system break down? A significant key to managing change is to make sure all the players are on the team with the responsibility to improve the supply chain or business. Without full commitment to the team, results will not be what are desired.

Agility has suffered from this random prioritization of change treatment. Success stories, like the one at Motorola, have been the exception and not the rule. As mentioned earlier, Dow Corning, a chemical company with its headquarters in the Midwest, was seeking to improve its competitiveness. In 1992 change was proposed by the organization's leaders. There would be two types of manufacturing processes inside of the supply chains. One would be the large-scale type of process. The other would be agile manufacturing, where the scale was not as large and batch operations were common. This was envisioned, defined, and communicated as a doctrine in what became know as the *Green Book*, which was an attempt to institutionalize the concepts. People were in complete agreement that this approach made sense. However, after four years, very little progress had been made in that direction. A few supply chains had picked up on the concept and saw the benefit that could result and change did occur, but a broad benefit did not materialize.

In analyzing the problem, several deficiencies could be detected. All the supply chain teams were not in place. The vision of the potential change did not involve all parts of the supply chains pursuing the agile direction. Teams did not focus on a supply chain team leader. Without a fully integrated effort to define how things would look after agility was incorporated into the supply chain, it was not clear how the benefit would be realized. How individual supply chain members would benefit was also unclear. Priority for change slipped in those parts of the supply chains that did not seem to reap the benefits but were required for overall success.

Another key reason for the failure to implement agility was because supply chains within the company lacked definition. Those that were successful had clearly defined structures with identified commercial leaders responsible for the operations. They had supply chain teams that were responsible for the performance of the business.

Despite the various problems connected with implementing change at Dow Corning, as improvement has been made in aligning resource and leadership toward supply chains, progress has been possible in incorporating agility into the change process. It does not mean that agility has taken on a life of its own. It is becoming a part of the overall improvement process of the various supply chains.

Necessities for Effective Change

At the 6th National Agility Conference held on March 4–6, 1997, in San Diego, futurist Joel Barker spoke on innovation and change. He concluded that inventors have the ideas, but an innovator is required to turn an idea into a product in the commercial world. This is probably the case with agility in modern industrial companies. It is at the idea stage and looking for innovators or change agents who will make it a part of the commercial world. Joel describes the 10 necessities of change as shown in Figure 2.1.

> **Figure 2.1   Necessities for Effective Change**
>
> 1. A perceived advantage of the change exists and can be identified by the user or person needing to make the change.
> 2. The change must be compatible with the thinking of the person making the change.
> 3. The change must be simple to understand.
> 4. The change must be divisible and doable in phases or by different people.
> 5. The change must be communicable and not involve a new vocabulary.
> 6. The change must be reversible and can be undone if it does not work as expected.
> 7. The change must have relative costliness. It must not require a lot of time nor can it result in a loss of face of the person doing the change.
> 8. The suggestor of the change must have creditability or a reputation of success.
> 9. The change must be creditable and do what it is claimed to be able to do.
> 10. The failure consequence of the change must be minimal.

Having all 10 elements established is desirable for a guaranteed successful change, but the rule is to have 7 out of 10 and always number 1 and number 10. In assessing many changes that were both successful and unsuccessful these rules have held true. Another absolutely essential element of innovation is that the change agent must treat the people involved in the change with respect. The Joel Barker doctrine of change makes sense. It's a set of rules that, if followed, will result in more successful change programs.

In addition to following the 10 key elements of change as described in Figure 2.1, significant change must be coordinated; that is, companies also must have a *process* that manages all the steps of innovation or change. This applies to shifts to mass customization as well as the incorporation of agile practices. The process starts with a vision being developed that includes agility and mass customization as key elements. To incorporate agility or mass customization into an organization's vision, the strategic thinkers must understand both elements and how they apply to the commercial situation of the firm. They must make agility and mass customization a part of their thinking so that they can be integrated appropriately into the supply chains. Since the concept of agility is not easy to grasp, the best way to understand it is to focus on both the concept, or tactic, of agility and the resulting benefit that will be derived from it.

Understanding How the Business Operates

A definition of business practices, or how a business operates, can lead to an understanding of how agility can be applied. This requires an assessment of how the business operates today and what it could be in the future. There is a long list of business practices that require thought and understanding before change can be initiated. Examples of ways to change existing business practices are shown in Figure 2.2. The figure defines how present practices can be improved.

The 10 concepts in Figure 2.2 are a few that need to be considered as change concepts are developed for a firm. They serve as examples of the kind of change that is required for improvement in competitiveness. The improved practice is developed with agility and mass customization in mind. At each step of building a new capability by altering business practices, unexpected events should be anticipated and reflected in the new capability. Economics become the dictator of what will be done to prepare for an unexpected event. The role of the supply chain leadership is to develop the business practices that will best fit the situation within which they do business.

Changing business practices is a part of a more complete change program that is improving on the activities within the supply chain. It is essential that those improvements and changes to business practices are coordinated and the efforts complement one another. The supply chain leadership and the team must define all the changes that are

> **Figure 2.2 Ways to Improve Business Practices**
>
> 1. Operate business as a *supply chain*.
> 2. Extend the supply chain to include *customers* and *suppliers*.
> 3. Operate the supply chain in a *continuous flow manufacturing* concept.
> 4. Minimize *inventory*.
> 5. Maximize *productivity* of all supply chain assets.
> 6. Operate the supply chain in a *just-in-time fashion*.
> 7. Institute a *make-to-order* capability.
> 8. Understand the *customers' real demand*.
> 9. Provide a high level of service by *grouping manufacturing capability*.
> 10. Shift capability to offer *custom products*.

desired and then prioritize them to determine what should be done first. In many cases, conventional improvements to the supply chain must be accomplished to set the stage for change in business practice. It is not a foregone conclusion that business practice improvement will be at the top of the list. In preparing for the prioritization process, the team must understand what the alternatives are in business practice and what change is essential to providing a reliable supply chain that can perform. A performing supply chain is mandatory in an improvement process. Once the priority is established, the implementation of the change process can begin. It must not be looked at as something that will happen quickly; it will be more like continuous improvement. The performing supply chain must consider all the capabilities and activities that affect the commercial outcome. It must be viewed as a business supply chain enterprise and not just a logistic one. This means it must include activities associated with material conversion to product, the flow and logistics of material, the information system that supports and directs the activities, and the cash cycle that brings in the revenue that covers the costs.

Developing a Measurement System

As change occurs in the various parts of the business supply chain and its activities, a measurement system must exist or be developed to show the benefit of various changes or improvements that are being made. The measurement system must also predict expected benefits and justify the consequences of the change. Performance measurement of the entire supply chain can be income statement related. It can also include return on assets (invested plant and working capital). It is critical that measurement include the cash cycle of the supply chain. The improvement process will be aimed at better use of installed assets, reduction of inventory, and an improved cash cycle. This is one of the key areas where benefits will be realized. Another key measurement is the growth rate of the supply chain; thus, the improvement process will be aimed at enhanced customer responsiveness. If this is done properly, then additional growth should be realized. All those who work on the supply chain must understand the benefit measurements. They must be the drivers of the change. This is a critical step in the improvement process. All benefits do not need to be at the income statement level but could be things that translate into performance measured at that level. An example of this would be reducing lead time in a supply chain by integration of information and the work process, thus reducing raw material, in-process, and finished-good inventories. Another change might deal with developing electronic commerce so that payments are made when the bar code at the customer's dock is scanned, indicating that the goods have been received and the truck is authorized to depart. Both of these reduce the investment that is made in the supply chain operation and indirectly drive the income statement toward the desired performance. The goals established in the implementation process need to be measured so that they can translate into real benefit for the supply chain.

Ensuring Success

Prioritized programs with goals do not ensure success. Nor does a clear vision with supporting strategy ensure success. The supply chain team must decide what is necessary to implement the program as well as ensure that the right level of effort is available to make the change. The team member who has primary responsibility for achieving the goal must also be identified. This is a critical step because the natural tendency is for people to resist change at this point. The task will look so large that all participants will feel the overload. Therefore, strong leadership is needed to help people realize that the whole change will require time and significant effort—a longer time and more effort than expected. The people responsible for change must have persistence and a will to overcome all obstacles. The change process cannot be looked at as a quick fix that will be finished in a flash but, rather, must be seen as an agile process that will frequently need redirection as the vision and the strategy are challenged. It must be viewed as a dynamic improvement goal to which the supply chain team is committed.

IMPLEMENTATION

With projects defined and prioritized, the tactics of implementation can be worked out. This is accomplished by those assigned change responsibilities. Since some of the changes will require changes in the jobs and responsibilities, it is paramount that people considerations play a key role in making decisions in this area. The impact of various improvement or changes in business practices must be understood. People should understand that jobs may be eliminated but the employees themselves will be cherished. Training will occur that provides team members with new skills. This consideration is essential if the change process is going to result in the improvements desired. This type of guarantee will help make the tactics of the change to business practice successful. The biggest impact of change will be on the operational supply chain.

THE ROLE OF AGILITY

How does this tie into agility or custom manufacturing? The supply chain of the future will operate in a much more agile fashion, with customization of product and service a key part of an enterprise's success. Properly incorporating agility and customization of product and services into the offerings of a firm or supply chain will increase the firm's ability to compete and grow. Although agility and customization are not the only business practices required for success, they become a part of the portfolio of change for the supply chain and how it will do business in the future.

Many activities make up the business supply chain. Changing these activities in an integrated fashion will change an organization's performance. For example, improved information and direction activity, like scheduling of the supply chain, can be combined with changes in the way materials are converted into products. To accomplish this, the business supply chain must become more streamlined and more responsive by minimizing inventory and shortening lead time. For the supply chain to operate with a minimum of inventory and a short lead time while providing the customer with the desired service, a number of changes must be made. For example, it may be appropriate to operate in a make-to-order fashion with delivery in three days from receipt of the order. Proper concepts will need to be worked out so this can happen. This will require the analysis of daily order patterns from the *customers*, not warehouses. This analysis may indicate that capacity will be exceeded several times each year and that an alternative solution to supply is necessary. The first reaction might be to buy more capacity and then let it stand idle much of the time. Another reaction might be to let the peak days of demand go unsatisfied. However, our vision stated that our goal is to improve the return on assets, so building more capacity does not seem like a solution. Likewise, not satisfying the demand will cause a distortion in the real demand as customers work to make sure they get the product, so this does not seem to be a solution either. This is where agility comes in to play. The capability must be developed to respond effectively to the unexpected demand. In the manufacturing plant a number of pieces of equipment look alike but are dedicated to make a limited product line for a focused set of customers. Analysis of the daily demand across all the products made by all the equipment shows that the demand could be satisfied 99.5 percent of the time if the different pieces of equipment were shared and scheduled together. Thus, by applying agility to the equipment and qualifying products on more than one piece of equipment, the demand could be satisfied without a capacity expansion. In the discrete parts business this can be accomplished by arranging the equipment in cells and instituting continuous flow manufacturing from those cells.

This same concept can be used in the batch process industry. Agility within the operation makes for enhanced profitability performance. Reduced investment in both inventory and capital assets provides the benefit that drives the effort toward the vision. This is one case of agile capability development that provides an agile solution to the unexpected change that impacts the enterprise. There are many other cases where agility will provide the solution to performance improvement. It is not just in the equipment capability but applies to all activity. It requires the understanding or anticipation of what unexpected change will impact each activity and predetermining an agile response to the unexpected event.

The visionaries and strategists in the company must understand what capability and options for improved business practices have been demonstrated either inside the company or in the outside world. This understanding can be obtained from benchmarking and study of successful firms. The importance of looking to other best practices cannot be overstated. Understanding what successful companies are doing and why should drive change activities in all companies. Even the best can learn from others. People who lead the corporation to significant change are usually the visionaries and strategists. They do not have the title "change leader" but operate from various positions within the corporation. Some are responsible for the future while others are involved in the day-to-day operations. It is the nature of the various individuals that makes them visionaries and strategists and thus the significant change agents. These people are continuously looking for good things that happen outside of their firm. They have a strong need to understand what others are doing and to make that a part of their change program. They thrive on finding something that has worked and can be adapted for use by their company. Using the concept of benchmarking outside adds validity to the ideas and thus makes them easier to accept. When this exists, the knowledge that the visionaries and strategists acquire from the best practices will find a place inside the enterprise or supply chain.

To drive change from external best practices, the leadership in a company must listen, learn, and then proceed to help others determine the direction that the organizational entity will need to go to become more effective and competitive. Without listening and learning, a concept like agility will not become a part of the strategy and thus cannot be included in the implementation phase. This applies to all new business practices and concepts, not just agility. Leaders of corporations must understand what the vision for the enterprise is and where they would like the business to end up so they can help people choose the right path to get there. It is an essential part of the change process.

Best Practices—Agile and Custom Manufacturing

To understand what agile and custom manufacturing can mean to an enterprise, it is useful to present a few examples. The intention is to provide some insight into how becoming agile can affect a firm. First, agility in an enterprise will impact the effectiveness of a supply chain by streamlining and better utilizing the investments that support the business. Second, it will allow the expansion of the business through a more custom product or service for customers.

Anderson Window, which will be described in more detail later in this chapter, presents a clear example of both benefits being realized. By improving the ability to easily accept an order for a custom window, Anderson was able to make the business grow. It is important to note that customers were looking for uniqueness in windows. They were ready for this option and supported it with orders. Anderson integrated the order system with a very agile shop floor that could respond to the customer request. The company viewed its shop floor as a process for making one window at a time starting with wood and glass. By developing this capability, Anderson opened a very large new business segment for the company. It was able to enter the market for the refurbishing of older homes and buildings where the window openings varied greatly in shape and size and were not historically standardized. Anderson's agile manufacturing capability supported the streamlining of the supply chain and the growth of two new businesses: custom windows for new construction and windows for refurbishing older structures. This is an example of a companywide positive shift to a more agile operation in a market where customers desired customization.

At Motorola, a shift to agile practice occurred in the weapons electronic systems part of their business. The customer was reducing its demand significantly and a change was necessary to make the electronic system product line. The customer was interested in ordering one system at a time and expected the quality of the electronic black box to remain very high. This was achieved by shifting from a nonagile dedicated series of lines to a single agile line. At the same time the supply chain was streamlined and lead times were significantly reduced. The new agile line was brought in at six sigma quality and it now produces weapons electronic systems on demand.

The U.S. automobile industry has a strong desire to increase its agility and to offer more custom options to the customer in a more effective fashion. A number of years ago the industry put agile practices in place. A regional and nationwide inventory system was put in place that would allow a dealer to find the car or truck that the customer wanted even if it were not on the lot. The car or truck could then be transported to the dealer and delivered to the customer and the sale would be complete. It used the 60 to 90 days of finished car inventory to satisfy the demand in a more universal fashion. However, this practice did not solve all the problems and fast-selling and hard-to-get models still require three months to be produced and delivered. In the automobile industry, the customer begins by looking for a custom car but becomes satisfied with what is available. The capability to add agility to the manufacturing part of the supply chain has not yet been available. The custom-designed car that can be delivered in less than a week is still an unsatisfied demand. Lean manufacturing prevails. The future may be a shift to the economics of a lean operation with an influx of mass customization. To accomplish this, agility must be effectively developed within the automobile supply chain.

Delphi, a supplier to the automobile industry, has developed a concept that allows seat fabric to be customized. A customer can specify the design of the seat that will be installed in the car that he or she ordered. The order system will maintain the integrity of the seat construction and ensure that it meets companies' standards of quality. The agile process will have a three-dimensional loom at its center. The loom is easily programmed to the unique or custom design. It also uses thread, not cloth, as the raw material. This would bypass a cut-and-sew part of the industry, so the supply chain would be streamlined. The concept sounds good, but the automobile people are reluctant to adopt it. Going from lean manufacturing to agile with customization is very difficult when viewed by enterprises as large as the automobile industry.

An example of both customization and a disciplined approach to agility comes from the aerospace industry. Boeing, when it set out to build the 777, worked with customers, operators, and users of the aircraft to develop and meet their requirements. It was a significant effort in customization within the context of the Boeing design-and-build capability. The plan was built to the jointly developed concepts and specifications. The design of aircraft was completely electronic; paper instructions, drawings, and specifications were not the means of directing either the design or production activity. Instead the task was performed by integrated computer networks working inside the company and with customers and suppliers. The airplane was designed by extending the supply chain forward and backward. It involved valuing the input into the design-and-build process of many more companies and their employees than ever before. Those companies were on both ends of the supply chain for aircraft. A network of suppliers from around the world build parts of the airplanes. The various components from tails to air-handling equipment plugged in without any need for adaptation. Watching the assembly of the plane in Everett, Washington, one sees workers using computer information for task assignments and technique descriptions. This can be contrasted with the 747 line where the system is on paper, and the instructions for assembly come from large books. Boeing built an agile system to support the custom design and set the standard for future airplanes. The Boeing vision included both agile and custom concepts as a key part of its overall improvement process. The company valued the judgment of the customer and the supplier in making the airplane better.

To achieve results in the preceding examples of customization and agility, the enterprises needed to understand where they were going. The development of vision for the firm was required and gaining support for the directions implied. The vision and strategy need to be well thought out and developed so that they can be understood by all participants. Once the vision and strategy are defined, understood, and supported, then the next step is for each team member and support person to define the projects needed to reach the vision. If the change is significant, then it is imperative to make sure that the time required to accomplish the end goal is understood. It may require progress that will occur over years. Even then, all the projects must be defined. Technologists must present changes that need to be made to the product or process. Market people will need to define how the relationships with the customer will need to change. Computer or information people will need to redefine the operational systems so they are consistent with the vision. In many cases the paradigm that puts a company on a monthly or yearly cycle will need to be challenged. These cycles are artificial and business is continuous, so that also must be considered when establishing capability. Other paradigms must be challenged and changed consistent with the direction, vision, and strategies.

With the projects defined, work can begin on implementation. Concurrent change needs to be coordinated so new concepts, business practices, or technology are available when they are needed. They must be working effectively when they are brought on-line. What comes first and what can wait until later is an important decision. The entire process must be driven by the vision and strategy. Establishing priority and time lines for each change is essential before programs are initiated. Once that is done, implementation can begin. Some of the changes will come on-line quickly, whereas others may require significant effort. Reaching six sigma quality or other standards does not occur overnight, and the program of improvement must be viewed as continuous. In some cases the process technology might not fit the vision, and time will be needed to develop a new way to make the product. Even a change in the way customers will be treated will require significant discussion and optimization. The implementation process is a significant effort.

During the process of implementation, it is essential to sit back and reassess. Do the projects and directions support the vision and strategy? It is important to look intently at the activity and understand what further learning is occurring and listen to the organization as it encounters obstacles and rough spots. Do these represent significant flaws in the direction, or are they something that can be overcome? The leader must listen, learn, and help maintain progress in the improvement direction. A keen eye must be kept out for flaws that change the vision or strategy either for the better or in compromise fashion.

Thus, the change process is not one where a leader can relax. Developing an effective vision and strategy to change a supply chain or company is essential to the company's success. Success must be planned and led. The process itself must be agile, with the ability to shift direction and steer the change.

INSIDE THE ANDERSON SUCCESS STORY

A more detailed look at a shift to a mass customization agile practice brings us back to Anderson Windows of Bayport, Minnesota. The founder, Hans Jacob Anderson, mass-produced window frames as far back as 1904. In the early 1980s the market was asking for more and unique windows. This became difficult for a company that had been mass-producing standard windows in large batches. Now, custom windows were required. The initial system was very difficult to implement. The price quote took several hours and then a 15-page specification needed to be developed. The error rate in this system was very high. The capability of the specification process needed to be improved and connected with the factory. In response to this problem a computer system was developed that allows salespeople to change and strip away features until they have the right window design. The computer automatically checks the window for structural soundness and manufacturability and then issues a price. Once the terms are accepted, the order is electronically transmitted to the factory where it is given a license plate and bar code.

The manufacturing process makes one window at a time, and the capability is agile. It uses many standard parts but then builds other parts of the custom window from scratch. This is a departure from the assembly line of mass production. Anderson's goal is to make all windows in batches of one with all components made to order. This pushes agile operations further into the plant. The custom order business was able to successfully reduce the error rate to 1 in 200 truckloads, although even more improvement is desired and errors are being traced to root causes and eliminated.

This approach has created new opportunities for Anderson in the window replacement market, where a new window needs to fit the space the old window was in. Custom windows are the answer. With custom windows, Anderson is pursuing this very large window replacement market with new technology and capability. The company is looking toward the future with agility and custom manufacturing as a very important element of its competitive posture.

As mentioned, the change to agility or mass customization is not always a road that can be traveled quickly. It requires persistence and commitment to a direction. It is a key requirement for the future and thus must be a part of every enterprise's change or improvement program. As in the cases of Motorola and Anderson, change must be implemented across the entire supply chain if the benefits of the shift to agility are to be realized. The benefit will be a more responsive and competitive company.

An Agile Strategic Planning Process

The concept that making change is hard work is demonstrated in a strategic planning and change process called *strategic mapping*. It is one of many such processes and will be used here as an example. Strategic mapping integrates the activities of strategic planning with all the stakeholders. It connects the thinking of the board of directors to the budget. It works at all levels and institutionalizes the direction of change. Strategic mapping is a process that is copyrighted by Geraci & Associates, Inc. of Pultneyville, New York. The concept was explained at a workshop at the Agility Forum in early 1997.

DEVELOPING A CHARTER

The process starts with an exercise with the board of directors of the firm, although it could also start with the leadership of a supply chain or an internal enterprise. The first step in the strategic mapping process is to develop a charter for the enterprise. This is done with the board of directors and the top manager of the firm. They essentially provide input on what they would like the firm to be. Things like "sustainable growth and profit," "ethics in business conduct," "reward and train employees to keep them the best," "operate globally," "increase financial performance," "gain market share," and "growth" are examples of goals. Developing a charter defines the field where the company will play. It's the focus, values, direction, expectations, and eminent challenge of the firm. This charter provides the foundation for the process. The members of the board of directors are individually asked to pick what they feel is the one most important thing. This provides a reality check and allows for prioritization of the inputs. Some of the listed things will not make the final charter.

The next step is to take the charter to the supply chain leadership. They will translate the charter into a vision. They will also establish items that will not be included. The process of defining a vision is specific to each area of the firm represented in the leadership—engineering, manufacturing, sales and marketing, finance, human resources, and procurement. These visions are all assembled and they can be very robust. Many people will want to participate in this phase, and they should be encouraged. All information is collected and assembled. It, with the charter, is input for the mission statement. The preceding discussion focuses on corporations with a functional organization or ones with a single supply chain. If the organization has functional people inside the supply chains, then it may be appropriate to bring the supply chain leadership into the process.

Creating a Mission Statement

The conversion of the charter to a mission statement requires the vision participants to break into teams and construct the statement following certain rules. The rules include a format with spaces to be filled in. The format is shown in Figure 2.3. This not an exercise in writing a pretty statement but is intended as one that can further define direction. The words should be as descriptive as possible.

The five spaces should be filled in as clearly and concisely as possible. The team should discuss and agree on the statement. The discussion will be very beneficial in future work on the direction of the firm. It should not be a limited debate but one that strives for unanimous consensus. Figure 2.4 gives an example of a mission statement developed in this fashion.

This exercise will result in a number of mission statements that then compete with each other. The various teams will be asked to cri-

---

Figure 2.3 Mission Statement Format and Rules

Fill in the blanks with candor and no attempt to wordsmith the meaning. This is a working mission statement that will need the real thoughts as a part of the work effort.

"_____1_____ is _____2_____ that does _____3_____ through (with/by) _____4_____ that results in _____5_____."

Guidelines to Fill the Spaces
1. Name of the firm or a new name that must last for two years or more
2. Defines active accomplishments
3. Function and deliverables
4. Strategies and alliances or unique property
5. The result for the firm. Such things as profitable and sustainable growth or ROI

> **Figure 2.4 Example of a Mission Statement**
>
> Custom Clothes for Women is a supplier to female customers of custom dresses, suits, blouses, and pants that are made by measuring, cutting or adapting, sewing, fitting, and selling through a network of franchised shops by people who are trained and equipped, with results in sustainable growth and acceptable profits in addition to happy customers.

tique the other teams' mission statement proposals by highlighting both the good and bad parts. This part of the process adds some stress to the exercise. This stress, however, will make the mission statement creditable. It will produce the definition of all the action elements that are important to the success of the firm. The mission statement will usually push beyond today's market and suggest new product offerings to give the firm or internal enterprise sustainable growth. Another example of the mission statement is shown in Figure 2.5.

This is not a statement that has been cleaned up for PR reasons. It deliberately has action words. The action words should be challenged as too broad or inconsistent with reality. The builders of the mission should be asked to defend words like *global* or *world's leading provider* or *multimedia*. They need to be defensible or should change to something more realistic.

> **Figure 2.5 Visual Images Mission Statement**
>
> Visual Images *is* the world's leading provider of multimedia solutions *that* design, develop, integrate, and support innovative systems for training, communications, and diagnostic applications *through* a highly trained, knowledgeable core workforce and reliable partnerships *that result in* sustainable profitability that meets stakeholders' expectations.

Transforming the Mission Statement into Goals

The next step is to change each one of the action words into goals. The teams would be reconstructed and go through a process of defining all possible goals in each of the action word areas, then narrowing them to five. Each goal must be measurable. If the team picks "reliable partnerships," they must define what this means as a goal, who will do it, and when. Once defined, the same thing is done for the other action words in the mission. This will result in a lot of information and directions. It is important for this step to take some time for audit and contemplation. Recycling and challenge are necessities. These goals become the basis for generating the strategies (how the goals will be reached). They will also define the imperatives. Goals must be purged, merged, linked, and then locked. The lock is getting the approval of the entity leader. The president or supply chain manager must agree that the imperatives are right. He or she must understand the mission, goals, and strategies that are the basis for the imperatives.

From Goals to Action Strategies

With the imperatives locked, operational planning actions are developed and put on time lines. These actions become the basis for the budgets that allow the execution and attainment of the goals. The overall process is the charter, the visions, the mission, followed by goals, strategies, and imperatives. This results in the operational actions and commitments that show up in the budgets. What looks like a long process is deliberately done swiftly to avoid too much detail too early and to ensure that only the most important directions result in action plans and budgets. This provides a focus.

Putting It All Together—Budgeting and Assessing

The strategic mapping process has a unique feature that makes it agile. Besides collecting all the good directions for the entity, it develops a budget. This step of the process looks different from what firms do today but the advance of computers and integrated operational systems will make this very easy and should result in more effective performance.

The budget process in this mapping has a horizon of 18 months, which is divided into thirds and reflects the tactical plan for the entity. It is 90 percent operational actions and 7 percent strategically driven with a small allowance for chaos and unforeseen events. Events in the first 6 months have a high degree of reliability. The next 12 months are a little more unsure, and less sure the further out one projects. The important thing is to have a budget that has actions that can be executed in both the operational and strategic change domains. The change domain is where the future will lie and it will support the strategic progress of the entity. There must be actions that make progress in today's activities.

Every three months progress is reviewed and the actions and budgets are adjusted. Every six months the budget is rolled forward and another set of actions is funded. This becomes a continuum. When one projects a few years in an entity, more and more of what is done is strategic and the near-term actions and budgets are preparing the firm or supply chain for the future. Of course, the knowledge that is obtained from looking that far ahead is not as accurate as the more near term. Continually looking forward and using the latest assessment of the future to retool the actions that will impact the future make this approach an agile strategic planning process.

Strategic mapping, developed by Frank Geraci, is one of many strategic planning processes. It is somewhat unique that in itself it is agile. It points out that developing direction and actions is a very hard job both for the operational and the strategic part of the firm. Instilling the desired direction change into the people who will need to make it happen does not occur easily; thus, participation in the process is essential. Deciding what agile practices will be developed inside a plan also is essential. The specific actions to achieve the benefits from having a more agile enterprise need to be measurable in the same fashion as other actions. It is at the action level where agility will show that it has merit and is worth developing as part of the way business is done. Leaders in the firm must understand what agility means and promote the concepts. Eventually it reaches the status of being part of a firm's culture.

This chapter has discussed the concept of strategy and change in an industrial world. The next chapter will give the background of industrial agility and bring the change process into the context of moving toward a more agile enterprise. It will introduce the concepts of business supply chains as a fundamental unit of commerce. It will briefly describe the three dimensions of new processes, products, and markets. It will also talk about building capability for agility for the unexpected changes that a business encounters.

PART 2

Operational Agility and Mass Customization

CHAPTER 3
Supply Chains

The operations of a company combine the key functions with satisfying the needs of the customers. Various concepts can be used to accomplish this integration. Function, geographic area or region, and industry focus have been ways to establish the integration structure. Another choice might be the business supply chain where all the resources and assets needed to satisfy the needs of the customer are aligned to a business supply chain management team. This provides the proper level of alignment of skills and makes sure that the customer is the focus of all members of the supply chain.

The following illustrates the growing importance of focusing on the supply chain. A report from the Manufacturing Roundtable dated October 1996, titled *Building Customer Partnerships as a Competitive Weapon: The Right Choice for Globalizing Competition?* (DeMeyer, Katayama, and Kim) describes the results of a biyearly survey that is carried out with 461 firms from Europe, Japan, and the United States. The purpose of the report is to gauge current thinking on global manufacturing strategies.

In section 8 of this report, the authors discuss the shift that is occurring in "Integrated Supply Chain Management" (ISCM) as a trend that has been developing and is currently considered very important in the eyes of the companies that participated in the survey. They describe, based on the information from the surveys, the four levels of ISCM. Initial efforts are aimed at rationalization of the supply chain network. The second stage is the sharing of information and problem solving. The process them moves to implementing systems. The most frequently cited systems were a total-cost approach and channelwide management of inventories. The fourth level is when sharing of risk and reward occurs, including tight partnerships. These four steps can occur with different intensity as the integrated supply chain gets more formalized and accepted. It is a significant trend for the companies surveyed.

Jay S. Kim brought together the information for a companion report titled *Search for a New Manufacturing Paradigm—Executive Summary of the 1996 US Manufacturing Futures Survey*. A significant portion of this report is called "Integrated Supply Chain Management—A Showcase for a New American Manufacturing Paradigm." The author cites the data from the survey to talk about the "shift of management focus" to ISCM. It is looked at as a tool to "transform the past improvements in manufacturing into strategic advantage." The degree of integration in the supply chain appears the highest in the following categories of flow: physical materials, information on current inventories, product specifications, and demand forecasting information.

Various categories of integrated supply chain initiatives, as seen by the companies in the survey, give some insight into what the future may hold. The top five of the nine future initiatives relative to what the survey respondents felt was important were as follows:

1. Channelwide management of inventory
2. Total-cost approach
3. Information sharing and monitoring
4. Joint planning and problem solving
5. Communication at multiple levels

Four others initiatives were listed as significant but not quite as important as the preceding five. They include reducing the distributor base, sharing risk and reward, extending the time horizon of planning, and reducing the supplier base. The nine future initiatives give an idea of what is important in supply chain management. The report also highlights the increase in interest that has been occurring relative to success or failure of the initiatives that a responding firm has in its portfolio. The supply chain, and the various initiatives that it encompasses, is gaining importance.

The supply chain, independent of the method of organization, needs to be the focus of the improvement process. Significant benefits can be gained in customer responsiveness and business profitability. Eliminating wasted time by improving the speed of the supply and logistic chain is the key direction for improvement. The overall goal will be to enhance the operational effectiveness of the firm. Another goal will be to improve the present operations of the firm to prepare it for an assessment of how business should be done in the future, which will address questions associated with the firm's competitiveness. Movement to mass customization, or make-to-order, may provide for significant improvement. Whatever direction a firm chooses, agility will play a key role.

Today's Business Organization

Many companies have emphasized functional expertise in marketing, which has required a liaison with the functional source of knowledge and direction that comes from a centralized marketing authority. This has been true for other activities also, such as sales, manufacturing, finance, quality, safety, and research and development. The balancing of power has been a key management principle that companies have used to attain results. Having that power aligned with the activities based on their expertise was a key way of structuring an operational organization. This approach relied on business activity to draw together the functional elements while the elements themselves paid allegiance to the function. Businesses needed to connect the "silos" of functional expertise to achieve commercial success.

Another approach to structuring the organization has been to emphasize the geographic area where a company does business. Companies would manage by designating areas such as Europe, the United

States, Japan, Australia, and so on. They would even divide these areas into discrete regions for management. This method emphasized the uniqueness or different degree of development of various parts of the world or regions. Within the global area or region, the organization was typically a functional organization. This concept combined knowledge of geographic uniqueness with expertise derived from the functional activity. The two joined to make a powerful link, and many enterprises thrived on this structure. The business activity was responsible for bringing both function and area or region together into a competitive force.

Industry focus has been another approach. In this method of operational organization, corporations would designate customers into industries so they could be serviced uniquely. This allowed the marketing, sales, and R&D organization to focus on an industry and its uniqueness. Products could be tailored to the industry as well as individual customer needs. The industry approach to sales and marketing could be maintained and the customer could be treated as an individual within the broader industry. The operational activities could adjust to the uniqueness of industry members and develop capability accordingly. The power of this organization was focused toward customers in the industry in which they reside. The internal activity then adapted its capability to the various customers and industries it served. Internally, the company operated with functional entities that represented the expertise needed to do business.

All of these approaches had positive attributes and also limitations. Even today, they are serving the commercial activities of many firms in an effective fashion. Different enterprises have picked different approaches to serving the customer. Adaptability to make the customer happy has been a trademark of industry. It has accounted for the success and competitiveness of many companies. Nevertheless, customers have been demanding more and more. They are asking for service that has previously not been required. Emphasis on profitability is driving firms to make significant improvements in their operations, which has put demands on the suppliers. Larry Bossidy who runs Allied Signal was quoted as saying:

> I'm asked all the time, "What worries you most?" And the most predominant worry is, are we going to be able to satisfy an ever more

demanding customer? I was talking to someone in the food business the other day. They deliver in ten-minute windows. In the aerospace industry you used to deliver in 30-day windows. Now it is overnight. Do we have the processes—the robust processes—to deliver in ten-minute windows? We sure don't have them today. Some of our competitors will have them some day, so we'd better be prepared. To meet these challenges, a company's people have to feel urgency and commitment.

Bossidy links the changing relationships with customers to the need to excel in order to compete. The firm must develop the responsiveness to meet customer expectations effectively. What was acceptable just a few years ago does not work today. Quick response is becoming the standard for doing business. Keeping up with the speed of the electronic communication media will require developing more robust capability. The process of getting the order and the paperwork done is shifting, with the help of integrated software and the computer, to the production of the product for the customer.

Many corporations have reassessed organizational structure and have selected new approaches. One approach is aimed at creating a responsiveness that can meet the customer needs of today and the future. This approach is called a *business supply chain*, and it involves structuring all the tools needed to respond into a given organization. It operates with an organization team and a leader who can bring the necessary resources to bear on the customers' needs to obtain results. The business supply chain becomes the power focus of the organization with cost, profit, and growth responsibility.

The Business Supply Chain

When looking at this customer-focused approach to operating a business, it is very appropriate to match customers with the supply chains that will best serve them. In this context it is both a business and logistic supply chain. The importance of having the business leadership interact synergistically with the supply chain and the customers it serves cannot be overstated. The business focus keeps the decision making at the supply chain level, with the supply chain business team and its leader, and provides the responsiveness for a true customer focus. The business and logistic supply chain contains all the capability needed to serve the customer and incorporates all the functional capabilities into an operating entity. The supply chain is a basic unit of structure within the corporation to serve those customers who are aligned with it. It has suppliers that serve it with the materials and components needed to do business. A supply chain that incorporates its customers and suppliers is referred to as the *extended supply chain*. Operational leadership of the supply chain has the responsibility to effectively make and deliver products to the customer and to operate the supply chain as a business entity. Customer satisfaction is the aim of the supply chain. It also collects revenue for the goods delivered and pays the suppliers for goods received. Because the supply chain needs to be as complete an entity as possible with profit-generating responsibility, it becomes the focal point for the improvement process. Vision, strategies, projects, and tactics are focused on this entity to add capability and improve performance. The team and leaders of the supply chain must determine how it will operate and do business. The three flows of the supply chain—material, information, and money—are shown in Figure 3.1. The supply chain cannot operate without all three flows working effectively. The team and leaders must ensure that the capability exists so that business can be effectively done.

Improving the capability of a supply chain is a critical element for improving competitiveness. How the customer is served and what return is acceptable are two major factors in directing the change programs of the supply chain; that is, the customer determines what improvements are important. In the Anderson Window example in Chapter 2 the customer was eager for custom windows. This initially occurred in new construction but was later used in the retrofit market. However, if the customer did not want custom windows, the Anderson strategy would not have worked. This was also the case at Motorola where the Defense Department supported the concept of downscaling to an agile response to customer orders. Clearly, the customer is critical in the determination of how a business supply chain will conduct busi-

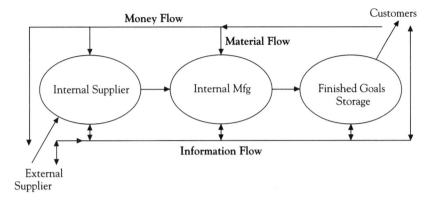

Figure 3.1   Three Flows in the Supply Chain

ness. This does not mean that the customer cannot be convinced that a new way will be a better way. However, if the customer is eager for the new product, then the effort to provide a customized capability is on firmer ground than if the customer must be convinced that it will be a significant improvement.

It is also important that the supply chain be defined and understood. Responsibilities and relationships, which are a critical part of a successful operation, must be understood, developed, and managed. Definition of the supply chain and the extended supply chain will enable the responsibilities and relationships to be developed and improved. The U.S. auto industry has done a reasonable job of defining its extended supply chain. This has occurred out of necessity as the industry focused on competing with the imports. It drove the first-tier suppliers to be more competitive and involved in improving their effectiveness. The improvement of the supply chain is where benefits are derived. Agility is necessary when a shift is made to a custom offering to the customer. It is also necessary as the supply chain is streamlined and actions are taken that enable the overall lead time of the supply chain to be reduced. Something that sounds simple, like "making to order" as opposed to "making to inventory," requires several changes, all of which bring agility into the supply chain. The economic benefits come from reduced lead time in the supply chain and thus a lower investment in inventory. These are significant and support the effort for change.

The U.S. automobile industry worked extensively with its supply chains and specifically first-tier suppliers. The industry provided the first-tier suppliers with industrial engineers to help them operate in an improved fashion. These engineers went into the suppliers' factories and brought world-class lean thinking into the supplier network. They worked with the suppliers to shift from production lines to cells. They taught the techniques of just-in-time delivery to the suppliers and also to the assembly plants. The auto industry also did extensive outsourcing to trusted suppliers that improved the economics of producing a vehicle. The entire industry changed. In that change the industry was working to drive the supply chain to a new level of responsiveness. Automakers wanted to be able to match production with the models and type of vehicles that were selling. The vehicle assembly lines were looking to change models as frequently as every two weeks. This meant that first- and second-tier suppliers needed to respond to this changing demand. The demand was driven by what was selling in the marketplace and what was in inventory. The suppliers were expected to match their output with the market variation. There was a continued push to identify what hampered this from happening and improve the situation.

One of the problems in changing output was fitting automobile interiors to the various model changes that were occurring because of customer buying patterns. When a model change occurs, the fabric used on the interior of the vehicle needs to be changed consistent with the model; thus, expensive models get better interiors than less expensive models. To change the fabric means that the yarn or thread from the textile industry must change in a consistent pattern with the model expected to be assembled in a few weeks. This change then results in a change in the woven fabric consistent with the need from the assembly factory. A new fabric can then be dyed and cut into the appropriate pattern. This is followed by the sewing into the shape for seats, ceilings, door panels, and carpets. This entire process needed to be accomplished in the two-week horizon of the shifting demand from the assembly plant.

In investigating the extended supply chain automakers found that the thread or yarn producer was operating from a concept of long runs with equipment working close to capacity. It was not agile enough to respond to the needs of the customer a number of steps down the supply chain. A solution to the problem needed to be found. The ideal solution would be for this step to be integrated with the change in vehicle demand. This would require a capability to make-to-order and delivery of the right thread to the supply chain in time to meet the changing model. The entire supply chain needed to change as the demand for vehicles changed. Thus, the entire supply chain hinged on the agility of the thread producer. The big question was whether the capability could be developed to perform in this fashion. It might require additional machines or the ability to shorten the conversion time from one thread to another. Finding the right solution and integrating the supply chain from the textile raw material to the vehicle delivered to the customer is a critical part of the improvement in the ability of the industry to compete. The industry spends more money on the textiles for the interiors of the vehicle than for iron and steel.

Success in this part, or any part, of the supply chain impacted the level of the inventory of finished vehicles that needed to be carried and the offering that was available to the customer. Balancing of the supply chain parts so that the whole can run with as short a lead time as possible is critical for improving effectiveness. Agility plays a key role in this activity. The quick change requires an agile culture, and an ability to respond must be a part of the capability of all parts of the supply chain.

The process of integration and balancing in the auto industry required an intimate knowledge of all the steps in the extended supply chain. The auto companies had the responsibility to build relationships that made all the tiers in the supply chain work in an integrated and balanced fashion.

Another example in this industry related to the inability of an assembly plant to get enough engines for one of the hot-selling sport utility vehicles. Not getting the engines meant loss in revenue and the assets or capability of the entire extended supply chain operating a little less efficiently because of reduced volume. This did not affect the model swings associated with demand or sales variation. It meant that the whole system would run slower. Once the situation was observed, a team from the assembly plant began to search for the problem. Engines had not been previously considered important in the supply chain. Those parts of the extended supply chain that had been improved because of previous problems were much better understood. Therefore, it took a month to track down the problem. After extensive analysis it was determined that the problem was the inability to get enough castings for a valve that went into the engine. Searching further found that it was the fabrication of the mold for the castings that was limiting. The molds were made from compacted sand with a binder. The binder was a clay material that had uses outside of the automobile industry. One of those uses was in the production of kitty litter. It seems the kitty litter people were willing to pay more for the binder than the foundry and thus got the material. It made for happy cats, but the customers of the utility vehicles were not being satisfied. The engine plants were not working to full capacity, because there were not enough engines for the assembly plants. Thus, the rest of the extended supply chain was slowed to the rate of the engine plants to keep production in balance. The definition of the problem and the solution took more than 30 days and had significant impact on the assembly and sales of the hot-moving utility vehicle.

This example points out to what extent a supply chain management must build relationships with its suppliers. This level of detail for a firm is critical to success. It may not have been critical when inventories existed between every step of the extended supply chain, but competition has made this a less than acceptable practice; today, the lean supply chain is the rule. Solutions to all problems and potential problems are required. The development of relationships that keep the whole supply chain striving for success is essential to the effective operation of the supplier networks. At each step the relationships promote the improvement process and drive toward a more responsive and agile capability. Again it can be seen that agility is a key and critical concept in providing capability and a quick reaction in a manufacturing environment.

The two preceding examples describe a look toward the suppliers. An equal level of analysis is needed in dealing with the supply chain forward to the customers. It extends farther than just understanding their needs for today and the future. Understanding how customers do business is also essential for the optimization of the supply chain.

Determination of how demand is managed helps you understand how to respond. In many cases the demand with the consumer or customer can be very steady and constant, but through the supply chain it can result in a whipsaw effect. That is, the supply chain can either carry large finished-goods inventories or have capability available far in excess of what would be required for an average demand. Enough understanding of the supply chain and the product demands must exist so that these circumstances can be avoided. Policies that are in place along the supply chain should make it operate efficiently. The supply chain team and leaders must determine how business will be done. Then through discussions and relationship building, partnerships can be developed that enable efficient operation. The focal point must be the supply chain and its customers and suppliers.

Supply Chain Improvement

To analyze a supply chain it is important to view it in terms of time and speed. In the supply chain, time is money. It is affected by the level of inventory investment. Something as simple as being able to change from one product or model to another quickly so that a make-to-order business practice can exist depends on the speed of change of the elements of the supply chain. In the computer world the concept of clock speed determines how fast things are running and changing. The same thinking is important with all elements of the supply chain.

The supply chain from supplier to the customer must be viewed relative to the movement of material. The value-added time—when work is being done on the material, component, or product—in a typical supply chain is surprisingly small. It is the active part of the manufacturing process and includes the movement of material from suppliers or to customers. It does not include the time that materials, components, or product sit doing nothing but being a part of inventory. Sitting and doing nothing is non-value-added time. Assessments would say that the value-added time as a percentage of the total time that material is in the supply chain could be as low as 0.5 to 5 percent. Time is wasted along the entire supply chain with inventory massing in interfaces between responsibilities. By managing the integrated and extended supply chain these pockets of inefficiency can be reduced. Inventory should exist only where it has a purpose. One of those purposes is agility improvement. Choosing where to keep inventory is a straightforward decision with relatively simple rules. One rule is that inventory should exist before the identity or final nature of the product is established. It should exist where a number of paths for further processing are available as the demand shifts.

Another rule is to push inventory back in the supply chain as far as possible. To do this the capability must exist to quickly convert the inventory into a product with final identity. Providing that capability is a part of the increase in agility of this part of the supply chain. Understanding where inventory should exist is critical to the effective supply chain. Deciding where the inventory should reside does not give permission to carry a large amount. The level must be established by the capability of the extended supply chain. A motivation for improving the supply chain capability is the reduction of this inventory and the increase in responsiveness. Agility comes from determining how the supply chain and its individual elements will operate and from understanding what the business practices will be that align with customer service.

A producer of silicone rubber in the Midwest operated in a make-to-order fashion with its customers where material was supplied within three days from receipt of the order. There were a few products where the relationship with the customer was a vendor-managed inventory, and the rubber compound was made and delivered to customers as they used it on a daily basis. The customers received custom manufactured products that were tailored to their needs. The products did not come from inventory but from a process that allowed the customization at the final steps of manufacture. Because of the nature of the product line, however, a large number of intermediate products were used to make the final compound. This created a significant inventory problem with some 150 intermediates feeding the custom order business. This gave the whole supply chain a third of a year of lead time. The value-added time as a percentage of the total was only 0.5 percent. This is well below the desired level of 25 to 35 percent value-added time of the total time in the supply chain. This was due to the machines that produced the intermediate products on a make-to-inventory basis and did not take direction from the make-to-order sales from the customers. The two operations were at different locations and operated as silos. Historically, the producers of the intermediates were the center of supply chain power, and they were not amenable to taking direction from the compound producer and the customer.

The solution was to schedule the intermediate production facility from the customer orders and make both the intermediate and the finished compound within the three-day order lead time. The integration of the elements into the supply chain resulted in a significant improvement in the speed with which the process and supply chain operated. These steps in the supply chain worked in continuous flow manufacturing with the customer order being scheduled and completely made without visiting the warehouse. Because of better use of working capital, a significant increase in profitability resulted. Further progress is possible and this product line will become a banner product line for the company. This example points out the power of supply chain integration with continuous flow manufacturing and the breaking down of the silos that stop people from working together.

It also points out a key measurement element for a supply chain: value-added time as a percentage of total lead time in the supply chain. The example showed that product or materials in the supply chain spent a predominant amount of their time sitting idle. When idle, with no work being done on them, they are not adding value. They are incurring cost. Besides the investment in inventory and storage space, costs are associated with moving material around in the warehouse and the workspace. Additionally, costs are associated with spoilage, damage, and obsolescence. To understand the supply chain, both idle time and value-added time must be a part of the analysis. Value-added time needs to be looked at to determine how activity will respond to unexpected change. This could be either product mix or level of production related. It is important that this activity be a prime focus of the agility or mass customization capability improvement. The value-added times are vital steps that make the capability what it is and determine much about how the supply chain operates. The equipment must fit with the expectation of the supply chain and be capable of responding in the way the supply chain demands.

Understanding why idle time exists is an important assessment process. The idle time, which can be 95 to 99.5 percent of the time, is where material sits and waits for something to be done to it. The concepts associated with continuous flow manufacturing are aimed at driving idle time to as low a level as possible. They involve the development of a manufacturing capability that finishes the production job once the product components hit the production floor. There are no shifts of material in and out of a warehouse. The production job is started and completely finished in a continuous flow fashion. The concepts of cells fit this definition with the making and assembling of the product all occurring in a cell location. The machines in the cell are balanced with the product lines expected to be produced.

It is also important to have the inventory where it can do the most good relative to serving the customer in an agile fashion. The agility concept ensures that the inventory is present for a purpose and not just sitting idle. For example, the location can be as finished goods, work in process, or as raw materials or components. Having it as finished goods usually means that the supply chain cannot be made as responsive as it needs to be.

Inventory probably should exist at a point where the identity of the product is not yet determined. It might be at a point where the product has not been pigmented or packaged or shaped. For example, a building sealant or caulk could be produced as clear or white and then flow in a continuous fashion to be pigmented as it is packaged. The packaging and pigmenting process would be driven in a make-to-order fashion with only the right amount of packaged material produced to satisfy the customer. This requires that the package supplier also respond to the order quickly with graphics for the package and that just enough packaged material be made to satisfy the customer order. This avoids any extra packaged and pigmented caulk going into a finished-good inventory that might move slowly. The intention of this agile capability would be to produce and distribute what is being sold in stores as it is being sold; that is, inventory would be responsive to the changing demand of the customer. Besides being agile in serving the customer demands, this technique can be used for customization. In customization, a customer might ask for private label graphics and a special color. The process would make the bulk caulk and then pigment and package the material to the customer's private label specifications. A portfolio of products with frequent changeover is possible. It starts from a bulk caulk and unlabeled packaging components and turns them agilely into goods on the retailer's shelf waiting to satisfy a customer need.

The preceding example, which is shown in Figure 3.2, demonstrates some significant changes that improved the supply chain. The monomers were received in a just-in-time fashion and were immediately polymerized as feeds for the compounder. The fillers were also received in a just-in-time fashion and were used immediately upon receipt in the compounder. The product was clear and waited to the last minute to be pigmented, and packaging indicated the custom manufacturing principle of postponing the determination of the product identity until the last possible minute. The flow through the process chain was continuous with no material ever going into the warehouse. Materials were processed through to finished product.

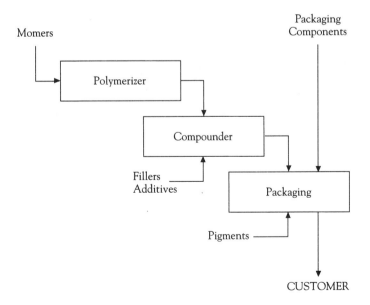

Figure 3.2 Building Sealant Process Chain

The building sealant process chain also shows the principle of supplier integration, where the graphics on the cartridge are put on and moved to the packaging machines just before the identity of the product is determined. No more cartridges are printed than the order dictates, and no surplus of either filled or unfilled cartridges goes to the warehouse. They all go directly to the customer in a continuous flow fashion. This process supply chain is a custom manufacturing unit with improvements that make it very effective. It is streamlined and delivers cheaper than process units that make to inventory. Custom manufacturing need not be more expensive.

In looking at the idle time in the supply chain, the first requirement is to eliminate it. This cannot be done just by asking for it to happen. It usually requires significant change. It might be that the firm needs to change the way it does business and then build the capability that facilitates the change. The change may be the shift from make-to-inventory to make-to-order. This change process would start by looking at the customer's daily order or use patterns. The real demand needs to be established by statistical analysis of customer order patterns. In the past these orders were satisfied by inventory in the warehouse. The desire is to have them satisfied by capability in the plant or factory. The information gotten from this assessment would then be analyzed to see how much capability is needed to receive, fulfill, and ship orders within a few days from the receipt. The exact amount of time used is determined by the customer's expectation. If enough capacity does exist to meet the peak order, then the assessment is whether the whole product line can be produced in an agile fashion. Variation in orders must be taken care of at an acceptable level throughout the year. The amount of capacity and its ability to respond will determine this. If the capability does not exist to satisfy the demand, then a strategy of shared assets might work. In many production environments, machines or equipment is dedicated to one or a few products. A number of machines might do the same thing. They may be limited by the ability to get the right materials to the machine to produce the complete product line. By qualifying the products on more than one machine or equipment, the surge in demand can be handled. It would be unlikely that demand would surge for all machines, customers, and product lines at the same time. The supply chain can be significantly changed with a reduction in finished-goods inventory. If done right, with inventory preserved at the right place in the supply chain, the whole chain can have a reduced lead time.

Assessing the use of time in the supply chain can make the supply chain more responsive and agile. It will also give a higher asset utilization; that is, inventory assets will be reduced, and the equipment can run closer to capacity. Agility can be managed, and it is essential to approach it from a supply chain perspective. The process of integrating and balancing all parts of the supply chain is essential to this improvement. It cannot be done in one element of the supply chain at a time but must be done across the whole supply chain. This further supports the need to understand the supply chain and to manage it as a fundamental business element.

Information is another important factor that can influence the idle time in a supply chain. It can be the information system that is not responding with the correct amount of quickness. Problems may occur with lining up transportation to move the goods to the customer, or delays may occur in notification of quality release or certification. Improper communication of what to make can cause significant problems. Loss of effectiveness may occur because the producer or plant may feel that it knows what it should be making, or it just may be producing what is convenient because of the lack of good information. Sales or marketpeople can persuade the plant to produce for its "best customer" instead of relying on predetermined priorities. The information system must be developed consistent with the needs of the supply chain. It must have the right level of capability and it must be disciplined. Accurate information and directions along the supply chain are essential for the whole to operate in an integrated and balanced fashion, which optimizes the satisfaction of customer needs and expectations.

Another cause of upsets in supply is failure—failure to make quality and shippable product because the technology is lacking and rework is needed; failure for the equipment to stay on stream in a predictable fashion; failure of the maintenance and production people to meet the schedule dictated by customer demand; failure to have the right materials or components at the right time, which can cause either short- or long-term upsets. The list of potential failures continues with as many reasons as there are activities. An agile supply chain must operate with disciplined and reliable performance. It is more important in an agile supply chain with short lead times than historic ones because an inventory of finished goods does not protect the operations. It is also critical that the information keep the status of machines, components, and product accurately and in all people's hands who are involved with the supply chain. If a failure occurs, and gets the proper attention, then it may be possible to find other alternatives and fulfill the customer's needs.

The concept of the "destiny of the supply chain" being in the hands of the team and team leaders is very important. They must understand what is presently happening along the operating supply chain. With this knowledge, the capability of the supply chain can be improved and changed. The changes should be aimed at reaching the vision and strategies that were developed to meet the expectations of the enterprise. Decisions cannot be made for one part of the supply chain without regard for how they will affect the remainder; the chain must be viewed as an integrated enterprise.

The decisions that are made will be aimed at increasing the supply chain capability. They will necessarily affect the various ways that business is done. Agility will be one of the significant considerations in the capability improvements as it is critical to improving the effectiveness of the supply chain.

This chapter has discussed the importance of supply chains. In the next chapter we will look at changes in the market, customers, and channels. This will be followed by a look at computers and information systems and how they make agility possible. This will be followed by a chapter on changing capabilities and automation for agility. These chapters together describe the direction of change in the operational extended supply chain—a supply chain that is agile but also integrated and balanced and can mass-customize its product line.

CHAPTER 4
The Extended Supply Chain

Operational agility is an essential part of the change process involving customers, markets, channels, and suppliers. They make up both the supply chain and the extended supply chain. This focus is an important part of the change process. Being responsive to the customer is essential for ensuring success. Customers are demanding a wider variety of goods and services that have been tailored to their requirements. Understanding the needs of the markets served requires intimate knowledge of individual customers. That knowledge is now influenced by global activity; consequently, the knowledge base must expand to cover the world. Few markets have only a regional limitation. Most have been developed to extend around the globe. For example, in the automobile industry the United States, Europe, and Japan were the major players, but now countries like Korea are becoming a factor and assembly of vehicles takes place in many countries.

Other industries have global markets. Textiles and clothing are made in many parts of the world for customers in other parts. Electronics are produced in Asia and sold in both the United States and Europe. The design of the product and management of the supply chain can be done in one country while the goods are produced in another. Being able to thrive in a global enterprise requires extended supply chain thinking. Just where the suppliers are located determines the cost and quality of the products that will be delivered. For example, a wholesaler or retailer of clothing might be required to get wool in New Zealand and have it made into cloth in another country, and made into clothes in even another country. Each step of the supply chain must be viewed for reliability, quality, and cost. The producer, wholesaler, or retailer must know who or what supplier company in each country will perform what step in the supply chain. The producer, wholesaler, or retailer provides the glue that holds the extended supply chain together. When well executed, the global extended supply chain provides high-quality goods to customers in various parts of the world.

Extending the Supply Chain

As mentioned, product quality needs have been increasing, and customers are continuing to become more demanding. Customers know that a wide variety of high-quality goods can be produced and delivered into the retail market at low cost. They expect the product to perform as promised and to be cost competitive. Making and delivering something that does not meet expectation is becoming more rare. Return policies of suppliers are liberal with "no questions asked" being a dominant feature. Trust in the consumer prevails or is required to provide the image that is desired.

The channels for supplying the goods and service customers expect are in a continual state of change. Shifts have occurred from the small store on Main Street USA to the large department store and to the discount stores such as Wal-Mart. The discount store is not restricted to general merchandise. For example, the home building and repair markets have companies like Builders Square, Home Depot, and Lowes. The office supply market has a significant discount segment. It does not end there but continues into appliances, CDs, books, computers, and other markets. The shift in retail product delivery has been dramatic, with non-value-added steps eliminated in the supply chain. The shift has occurred with the customer benefiting from higher-quality products at lower prices.

The shift to the discount store is not the only change in retail channels. Another shift is to the high-quality specialty stores that aim themselves at specific niche customers and then provide the variety, cost, and quality that these customers are looking for. Specialty stores deliver products that have the reputation of performing better than those that are available from department or discount stores. The type of merchandise ranges from products in specialty sporting goods to those in nature stores, such as for bird watchers. The large variety of products that is available ensures that the consumer can find the right product.

Another shift in retail channels is to the increase in catalog sales, with many people choosing to pay a little more for the convenience of doing business from home. This does not apply only to the U.S. market. Many catalog sales are in countries like Japan where goods like outdoor and trendy clothing of high quality can be gotten at low prices by ordering from the United States. The genesis of catalog sales began with companies like Sears and J.C. Penny, which had department store catalog sales for many years. This has expanded to the point where catalogs are commonplace in the daily mail delivery across the United States. Conversely, many of the companies that built a clientele from catalogs opened retail outlets to become more available to the customers. This provides another step in the changing world of retail sales.

These examples show how distribution channels are everchanging and the free enterprise system provides a rapid test for a channel's effectiveness. With consumers demanding high quality and competitive prices, little tolerance exists for the delivery of products that do not meet customers' needs. Free enterprise has made this part of the distribution supply chain a very agile activity, where a large variety of goods is offered in a cost-effective fashion and new approaches are developed constantly.

The improvement in the distribution of goods through channels is not restricted to the consumer market. Distribution channels between one manufacturing firm and another are also being streamlined. The value of elements in the channel is continuously being questioned. Electronic communication has changed the information flow associated with both the consumer and the industrial distribution of goods, services, and cash. For example, the electronic transfer of payment is becoming a part of doing business. Capabilities within the supply chain are changing at a rapid pace to keep up with the change in distribution and market needs. This change will undoubtedly continue into the foreseeable future, driven by increasing capability and the flow of information along the entire supply chain, especially with customers.

There is an obstacle to an integrated information and cash payment system across the extended supply chain. It deals with the offerings of the various providers of integrated information systems—sometimes called enterprise resource planning (ERP) systems—like SAP, Oracle, PeopleSoft, and JD Edwards. All these products do a good job of integrating information systems into a company's internal supply chain, but they do not address the question of how the extended supply chain functions when different software is used for the internal integration. The systems cannot talk with each other. Thus, they need a common data storage system with the right kind of security so that firms can work with each other. This will enable firms on a supply chain to do electronic-based planning and cash payments. Electronic commerce will benefit greatly from the software provider that develops a solution to this problem. The need is very large and an "extended supply chain server" could be on the horizon. Currently, however, the ERP software providers are gaining exclusivity from not being compatible. This lack will hinder collaboration between trading partners for continuous replenishment, cash payment, and transportation management. It will be a constraint on effectiveness and agile operations.

Determination of how the extended supply chain will be developed is critical in the definition of how business will be done. Extending the supply chain forward to the customer can involve using distributors and retailers to make product available to the customer. Extending the supply chain back to suppliers and contract manufacturers in the supply end of the chain is also an important consideration. Extending the supply chain in either direction involves assessing the degree of vertical integration that is appropriate for the particular supply chain under consideration. The enterprise may choose to supply all the raw materials and components for the production process if these are a critical capability for the firm and determine the competitive advantage of the supply chain. If the raw materials or components are not critical to the supply chain success with multiple sources available, then they can easily be obtained from others that have competency in their supply.

The degree of vertical integration forward toward the customer must also be determined. Using distributors to get the product to the customer is a possible choice. This could involve a part or all of the product line. The distributor could provide service to the customer, or another firm more capable in the market could perform this activity. An enterprise has many choices about how to do business in the market. These choices rest with the supply chain leader and the team and will determine the nature of the enterprise.

The decisions on the role for suppliers, channels, and distributors must be made in a disciplined fashion. They must get serious consideration since the nature of the enterprise or supply chain is at stake. Just what portion of the supply decision is put in the hands of those outside the enterprise is a function of where the enterprise feels it has core competencies and where it gets its competitiveness from. There are "make or buy" decisions along the extended supply chain, and the degree of vertical integration is a result of those decisions.

Customers, Markets, and Channels

The capability to make and deliver product and services is in a rapid state of change consistent with the change with the customer and markets. Modern consumer product distributors like Wal-Mart are expecting suppliers to take over more and more of the activity of keeping goods on retail shelves. This trend is beginning and will likely exceed what has been happening for a long time in the grocery business. The suppliers of bread, soft drinks, snack foods, and other products have been detailing shelves for a few decades. In today's competitive world the major discount retailers provide cash register receipts by location by item, which then become available to suppliers to restock shelves at individual retail outlets. This has presented a significant challenge. For example, a supplier of building sealant is faced with having 15 to 30 items on a store shelf to satisfy the demand for different colors and performance. The supplier monitors the cash register receipts and then must provide the right quantity of material to restock the shelves. This must be done at the right time with the right level of agility. Some materials will move quickly and others slowly. Minimum delivery might be a dozen packages of any one material. The self-generated restock order must fit the need of the individual store and needs to be assembled as a unit for delivery. At the factory this could be assembled from inventory or it could be made to order. It also could be a combination of the two approaches. Regardless of the approach, the retailer expects the shelves to be stocked or the products will be replaced with a competitor's offering. Firms are presented with a real challenge as change requires a more customized service.

The make-to-stock approach appears to be the easiest for a manufacturer to manage. The products are stocked in the warehouse and delivery occurs from this inventory. As product is shipped, a reorder point is reached and the manufacturer schedules more to be produced. These are classical material requirements planning (MRP) activities. The minimum stock level in the warehouse must be large enough to take into account the demand for the product and the capability or agility of the manufacturing plant. It must understand how quickly and frequently it can shift from one product to another and how long restocking will take. Less agility means more inventory when the replenishment is scheduled at the reorder point. The stock level is set by the analysis of the demand and the ability of the manufacturing to respond.

The disadvantage to this approach comes from the level of inventory that must be held and the cost of carrying it. In the example in the previous paragraph the building sealant product is typically privately labeled. This adds complexity since a dozen or more retailers are expecting the same private label service with package graphics designed for their retail offering. The warehouse would need to keep inventory levels for all the graphics; this inventory can be very large, and the warehousing activity can be extensive.

Another choice does exist to satisfy the retailer's demands. It involves building a more agile capability in both the building sealant and package manufacturing process so that it can respond in a make-to-order fashion. This will require building an agile response in both segments of the business. In this situation one could make a single color of sealant such as clear. This material would flow to the packaging lines where the orders are being satisfied in a make-to-order fashion. The pigmenting of the sealant and the printing of graphics would be done in unison at the latest possible point to feed the packaging machine. Just enough material would be pigmented and packaged to satisfy the orders. It would be boxed and assembled on the pallets with other material from other lines to fill the order on the day it was received. In this case the order is generated inside the firm from the cash register receipts. It then goes onto a truck that takes it either to the store or to a redistribution center where the retailer will move the pallet of material to the appropriate store. All the material pigmented and packaged is shipped to the retailer. The quantity produced exactly equals what has been ordered, so nothing is put in the warehouse. The minimum order size might be a pallet load of material and the minimum color might be 10 cases.

When the goods are shipped, the invoice process starts. When they are received, they are scanned and the bar code read. The payment process can then begin. The scanning also allows the shipper to depart for other delivery or return. The terms of payment are negotiated ahead of time and the computer will take the appropriate action. The trucker will leave with the scanner informing that receipt has occurred. This will be transmitted to the supplier's computer system, where it will wait for acknowledgment that the goods were received and that electronic transfer of payment is occurring. This approach is very desirable but is limited today by the integratability of the extended supply chain and the absence of an information data bank into which the customers and supplier can input data as well as access it.

The retailers are working on an improvement, or a new twist, to this process. They would like to use the cash register electronic signal to indicate that the goods have been sold and that payment to the supplier is warranted. This makes the supplier responsible for miscodes, theft, damage, and slow sales. This approach would take retailers out of owning the merchandise and keep their cash investment at the lowest possible level. The manufacturer would carry the cash cycle and have the responsibility to provide product that moves and to not overstock the shelves. This improvement or twist is not yet in place, probably because suppliers are not comfortable with handling the massive amount of data that would need to be processed continuously in an error-free fashion. It is also true that suppliers would not be eager for this to happen unless the system and logistics were very reliable. Although eventually the capability will exist to move both the materials and the information effectively, this improvement in the distribution channels will need to be reliable enough to work. Manufacturers or their distributors in concert with the retail firms will need to build the capability to make this happen effectively.

Retailers' demands can be satisfied in another way. The cash register receipts can be monitored to determine the demand on a daily or weekly basis. The enterprise can make to order as the demand is occurring. The order can be assembled with slower-moving items taken from inventory while the faster movers are made to order. This requires an agile capability for the fast-moving materials. As in the preceding make-to-order situation, integration of the package graphic capability with the pigmenting at the point of packaging is required. The demand would come from orders based on cash register receipts. This method would also require an inventory of the slower-moving material that can be assembled on the pallets with the faster-moving material and be aimed at delivery at a particular store. The whole system must be agile. It does not happen by accident but requires the building of a capability. The benefit of delivery by a make-to-order capability for the supplier is not having as much cash tied up in inventory. This example describes the changes that are occurring in the retail marketplace. They require the changing of the capability of firms along the supply chain. They must operate in a streamlined fashion to minimize cost while maximizing responsiveness and reliability of supply. The most competitive firms will be those that deliver product in the most effective fashion with the least cost.

Clearly, the way companies do business is changing at a very fast rate, and the speed of change is driven by the large benefit that can result. The benefit comes from the improvement in the supply chain, with each step being assessed for the value it brings, and responsiveness to the customer. Additionally, the overall integration of the steps so they work in unison is being accomplished and significant benefits are resulting.

One benefit is reduced inventory. Carrying inventory is expensive. Estimates of the cost are from 25 to 40 percent of the cost of the goods. The cost includes the carrying cost of money, warehouse and equipment, breakage or damage, and obsolescence. Because the cost of carrying inventory is so high, this facet of the supply chain is something that is getting prudent supply chain management's attention. It is a place where innovation is occurring. Much of that innovation involves becoming more agile in all the activity in the make and deliver cycle to create a significant competitive advantage.

The second benefit of change in supply chains comes from responsiveness to the customer. A strategy of customization of the product to the customer can result in increased growth and provide an exclusive or proprietary right to the market. To operate with a custom product offering, the customer and market must want it. This means that there must be a value for both the enterprise and the customer in having a custom-manufactured product. The enterprise can improve the supply chain and can gain market share by more capably satisfying customers' requirements. The channels must be capable of handling this way of doing business, and the manufacturer must be agile enough to provide the product in an effective fashion. Even custom products must operate with a lean and streamlined supply chain.

One of the ways to measure the effectiveness of a supply chain is by looking at how long it takes to go from raw material to finished good to the retailer or customer. This time is normally expressed in days. For example, the chemical industry has about 80 days of inventory using industrywide numbers for the United States. This level in a process industry is often viewed as chemicals flowing in pipes, being processed, and then shipped by rail tank car or a tank truck. This continuous flow model of the chemical industry is a misconception, however, because the industry gets much of its production from batch equipment, and shipments are in containers smaller than the tank truck. This batch operation can result in large in-process and finished-good inventories unless they are managed in an effective fashion.

Much can be done to improve the management of the chemical industry's batch operation, supply chain by supply chain. An example here will illustrate. A firm located in the Midwest brings a basic chemical in barges from the Gulf Coast. The chemical goes through extensive processing and is tailored or customized to individual customer needs. The firm has 500 different products, produced in five different plants located in the Midwest, that are made to order for many customers by altering the way the chemical processes operate. The customers demand that their product be delivered in 30-minute windows of delivery in a just-in-time fashion. The package containers are rail tank cars, tank trucks, and drums. After an extensive effort to change the way business had been done in the supply chain, a significant improvement resulted. The lead time from receipt of the barge by the firm until delivery to the customer, in that 30-minute window, was reduced to 15 days from what had been close to 80 days. This reduction took a couple of years to perfect, and some key changes in the way business was done needed to be made. Very few of the changes were in the physical assets of the firm. Most were in the concept and relationships of the way business had been done.

One of the first things that was accomplished was to devise a realistic way to treat customers. After a long period of discussion and negotiation, a system was established that put all customers into three categories. The one-star customers were the best and the ones who the supply chain would do anything to satisfy. The two-star customers were ones who would get almost anything done for them. The three-star customers were distributors and had lower priority. To make this system work, all business in containers smaller than a tank truck was given to the distributor network. This meant that distributors would supply the smaller container needs but not rail or tank truck business. The distributors realized that their orders would not be filled immediately from the plant and had to plan for inventory until the product they wanted was scheduled for production and made.

Another important change was to get the plant managers from the five plants to make changes the central demand planners said needed to be made. This was a make-to-order business, and the plant management felt that they could predict the demand and customer orders better than the people looking at the whole demand. However, this approach had resulted in inventory and the inability to use different plants for different customers in an optimized fashion. The order entry people became the supply contact for the customers and they were able to take the order and find out about when another would be needed. They captured the pulse of the business by building relationships with the schedulers on the floor of the customer. It was no longer necessary for customers to be visited by salespeople to establish the demand. A number of handoffs of information were eliminated. Even with this improvement the resistance to this centralized approach was high, and the final plant manager had to be convinced that it was a condition of employment to follow the centrally issued production and delivery schedule. One month after the condition of employment ultimatum, the plant manager who had initially resisted the change was a strong supporter of the system. This change gave the firm the ability to use its assets in an agile fashion while satisfying its customers.

The change took time. The marketing people needed convincing that dividing the customers into three categories was a good idea. They also needed to be convinced that they would no longer be able to influence the delivery schedule by influencing the plant manager. They had to redefine their relationship with each of the customers. The plant managers likewise needed to honor the three classifications of customers. They needed to understand what "do anything" meant. They also needed to follow the centrally generated make-and-deliver schedule. The plant managers needed to overcome the temptation to make what was the easiest based on the raw materials that were available. They needed to be a part of the whole supply chain system and help it work effectively to satisfy the customers. They needed to let the central system manage the crises with all the alternatives available to them.

The central planners had a procedure that was important to the success. When taking an order, they would ask the customer when the next order delivery would be needed on the products the customer received. This gave them a look into the future that was much improved from previous practices. The activity of order entry and fulfillment was occurring at the base level in the customer and supplier locations. This in itself built relationships based on trust and real knowledge. As mentioned earlier, the changes improved the effectiveness and allowed the reduction of supply chain lead time to less than 15 days. This company's experiences illustrate a real success story and a study in the application of agility to achieve success.

Agility in today's supply chains is also becoming evident in the role of the distributor. Historically, distributors have provided the sales force to a given class of customers. They handle a portfolio of products from different suppliers and represent these products to their customers. They also take orders from these customers. They arrange transportation and deliver goods from inventories that they hold. They manage the receivable from the customers and balance that with the payables from their suppliers of product. They are a very key part of the supply chain. For example, in the preceding example, a role was defined for distributors. They were given the task of handling the drums and smaller containers. This meant they needed to understand that business and provide the needed level of inventory to satisfy their customers.

The role of the distributor is in a state of change. It is being challenged by the new technology of the information age. The information age enables firms to deal directly with customers of a wide variety of sizes. Trucking firms can ship directly and economically to customers because of improved information systems and on-line communication and direction for the truck fleet. More goods are moving through high-volume outlets. The growth in catalog sales and discount stores is presenting a different situation for what has been known as distributor channels. Agility by various people along the extended supply chain challenges the role of the distributors. The concept of make-to-order and the building of the capability to respond to that type of business will challenge the ability of a distributor to respond. Just what role distributors will have in the future is an open question. Perhaps they will serve as suppliers of outsourcing expertise in the context of the supply chain. This expertise will enable the delivery of specific skills and materials to companies so that the firm and the supply chain can operate efficiently. Other alternatives are also possible. Whatever the role, it will require competing in a more competitive world.

Suppliers

Suppliers are an essential part of the supply chain. They provide the materials and services that an enterprise needs to be successful. The relationship between a firm and its suppliers must be developed so that it is consistent with the importance of the material or service. If it is a critical one, then the relationship must be strong and based on trust. If it is something that is available from many suppliers, then the relationship need not be developed as fully. It is not only materials that require the development of appropriate relationships. People are also involved; for example, someone like a logistics or transportation provider must be looked at to determine what type of relationship is the right one. Other relationships that involve "make or buy" decisions truly determine how a firm would like to do business and where core competencies are to be developed. These decisions deal with the vertical integration of the enterprise and are fundamental to how a firm will conduct itself.

In recent times enterprises have been questioning what they feel they must do themselves and what can be done by suppliers. They have determined that others are more proficient in many things that historically have been done internally. This involves the establishing of suppliers that will provide the service or material. The initial approach is to outsource functions that are not critical to the firm's success, like building maintenance and cafeterias. Outsourcing can be expanded to computer service and payroll. Currently, firms are considering everything a firm does as a candidate for buying the service, product, or raw material.

A frequently asked question is, How does a firm determine when buying is better than making or providing? The decision process to determine the nature of an enterprise's supply chain should start with a determination of the firm's core competencies. Core competencies are what make a firm what it is or what it might like to be. They are unique to the firm and they should be the something that they have that few others have. They should not be confused with things at which the firm is competent. Some might consider McDonald's core competency to be making Big Macs. The company's assessment probably says that its core competency is the promotion of fast-food sales. It also thrives on a consistent and predictable food offering at a low price that is delivered quickly and has a taste that people enjoy. It takes a lot of things that an enterprise is *competent at* to make the core competencies the foundation for success. With the competencies understood, the firm can decide which are candidates for having someone else do and which must be retained inside the firm.

A firm that assembles, integrates, installs, and services communication systems would not look at the manufacture of the components as something essential for them to do internally. Their expertise is putting the right systems together and making them work. They can get the components from the market or have someone custom produce parts that might be unique to the integrated system. The core competency of this type of firm is the intellectual property associated with components operating in an integrated fashion, providing customers with visual and audio communication systems on a global basis. It is not in making the components.

Other firms that are their best at producing something and are the best in the world at doing it have a core competency that relates to the manufacture of products. This may not be the only core competency, and many enterprises consider the introduction of new products as a foundation core competency. Neither of these would be candidates for having someone else produce or introduce products. Most everything else would be a candidate. Indeed, the marketing and sales of the product could be turned over to distributors.

Making the decision on when an enterprise will have someone else make a product or perform a service follows the rule that the core competencies of the firm are not candidates for the outsourcing decision. This can change as the market and the competition changes and should not be an unalterable decision. With core competencies understood, all other activities can be considered for outsourcing.

The global automobile industry has been through significant change relative to what it does internally and what is outsourced to suppliers. Comparing Toyota with the typical U.S. automobile company shows that they have outsourced many more of the components that go into a car than have U.S. producers. U.S. companies were organized initially to produce everything needed for the automobile. Suppliers did not exist and almost everything was done inside the company. Glass and steel were produced by the Ford companies. They were almost a stand-alone enterprise. Toyota had a much different experience and developed component suppliers that produced more than 70 percent of what went into a car. General Motors did almost the opposite, with 70 percent of the parts and components sourced internally. This probably reflects what each firm believes is its core competency and the historic basis for the firm. Toyota has stressed the making of the machines that are used for production. They feel that one of their competencies is the infrastructure and management systems used in operating the enterprise. GM buys the machines and infrastructure to produce the car and focuses on the act of manufacture as a competency. Toyota's focus on the machines and system has resulted in the lean manufacturing system. Infrastructure is where the technology is being developed and exists to improve the design and manufacturing processes. U.S. firms in the auto industry are typically lean in the infrastructure and not as lean in the design and manufacturing activity. Japanese firms developed the infrastructure that has enabled them to control the improvement processes of both design and manufacturing. Both centers of automobile production have stressed the technology, with one focused on advancing it while the other focused on using it. This has changed over the years, and the two are closer together today with U.S. firms working to improve their systems and to become leaner. Competition in this industry has been and will continue to be at a very high level.

When the U.S. Congress funded Sematech, AMTEX, and the Agile Forum, they were concentrating effort on the development of technology to produce. They were focused not on the act of making a product but on the technology, infrastructure, and machines that make the product more effectively. It was the industrial infrastructure that would make the industry more competitive. Sematech has shown that this process can work, and the electronic industry leadership shifted back to the United States. The participating firms in the Sematech activities have been able to compete in ways not thought possible. The lesson is that a core competency must be the technology to produce or offer a service that will enable firms to improve their effectiveness.

Summary of the Extended Supply Chain

The extended supply chain is key in effectively executing business. The effectiveness of the total operation from the suppliers to the customers must be the focus of the improvement process. The strategies and business practices must be compatible with customers, markets, channels, and suppliers. One cannot do business differently from what the extended part of the supply chain desires. If the customer does not want a customized product, it would be foolish to offer it.

Much change is occurring with customers, and a supplier must be in step with those changes. Retailer business continues to shift toward large volume outlets and catalog sales. These customers are more demanding and are looking to have vendors detail shelves and manage the inventory in the store. The catalog people are developing large volume and it is important to be able to respond to them in a make-to-order fashion.

Private labeling and specialty shops offer another unique challenge. The customer demands high-quality goods effectively delivered and at a reasonable price. The future holds many additional challenges in improving the effectiveness of servicing the customer.

If suppliers can't perform effectively with just-in-time delivery, then something different must be worked out. These considerations are paramount in considering the way the supply chain will do business. Agility and effectiveness improvements must be consistent with their capability.

This does not mean that a program cannot be developed and executed that changes the entire or part of the extended supply chain. The Ford Motor Company did it with its supplier network. It set the standards of performance in both quality and productivity and then helped the supplier meet that standard. But it was not a short-term effort. It required a lot of energy and work. Suppliers were challenged and Ford's motivation was questioned. The effort went on for a considerable time and many changes were made, all directed at suppliers and at supply chain effectiveness improvement.

Another issue in working with the extended supply chain is the determination of what part should be in the extended area and what should be in the internal supply chain. The commercial world has changed recently, with many firms going through make or buy decisions. This process deals with both services provided to the firm and the manufacturing and delivery process. In assessing how business is done, this becomes a critical exercise, where the choices can have a profound impact on the extended supply chain's effectiveness. Looking outside the firm to see what service or manufacturing capability could be bought as opposed to conducted within raises new possibilities and flexibility. Outsourcing is a process where significant agility can be acquired using external capability that is owned by others. The make or buy decision should not be treated casually, since it is fundamental to how a firm will do business.

This decision should be looked at carefully with the supply chain and the enterprise defining its core competencies and also other activities where competency exists. This provides the framework for outsourcing considerations. A firm should not consider outsourcing what makes it unique or is a core competency, but everything else should be a candidate. The decision involves finding capability outside that is more effective than what exists internally. In many things this is easy to do and the economics will point toward shutting down the internal capability and having it done externally. However, the decision is not entirely economic. Improvement in responsiveness and agility must be considered. The outsourcing decision is usually a part of an overall improvement process, and the desire to change the way business is done is a major factor. With the new degree of freedom offered by others outside the firm, it is possible to build a much more agile extended supply chain while changing the business practices. The process requires developing the understanding of where the firm or supply chain is headed with regard to the new business practices and then finding the right capability outside to execute. The ability to respond to the unexpected but anticipated change must be a part of this exercise.

The total extended supply chain should be included in the improvement process. A lot can be done to bring agile capability into the extended supply chain by integrating external capability with internal. Having the freedom to look at outsourcing in the broadest sense is a requirement for finding those capabilities that are synergistic and reinforcing to ones that are to be kept internal. The focus should be on doing what one does best internally and then selectively and smartly building external capability that makes the whole supply chain more agile and more efficient.

This chapter has defined the need for change and an agile response with customers, markets, channels, and suppliers. It also has highlighted the choices that a firm has in dealing with its suppliers and what can be outsourced. The next chapter deals with the role of the computer and information within the agile enterprise and the extended supply chain.

CHAPTER 5
Integrated Information Systems

A new age is under way, which has led to a dramatic decline in the cost of computation and storage over the last 10 years. The advance of the PC, both laptop and desktop, has dramatically changed the world in which we live and work. An initial result of computer technology has been to change the work that we do. It has been the primary reason that productivity of the white collar world has increased. This increase in productivity is one of the reasons that reengineering of firms and the displacement of people from jobs has occurred. Increased productivity is not the only reason but, combined with outsourcing work, it accounts for much of the change in the number of people needed to execute business. The new age has been accompanied by the development of software that fits the needs of many firms. Firms specializing in specific software can be developed to a more robust state. Specialized software, plus the integration of data systems, provides the typical firm with a much more robust system than has been possible in the past. The expectation is that software firms will continue to perfect and expand the capability of their products. Improvement in the availability of information will continue and automation of decisions will be possible. The results will be a continuing increase in productivity and better decision making.

The typical manufacturing company no longer develops software for itself. Rather, it uses others' software packages. Because software packages have logic systems built into them, they will enable the same decisions to be made in the same fashion under the same circumstances. Placing routine decisions into a systems-based process will remove them from becoming a repetitive activity. The focus of the operating people will shift to the unique decision and expanding capability. This chapter will deal with manufacturing process information systems, integrated product and process data systems, process control systems, the integrated enterprise system with modules, and scientific computing. All of these technological capabilities are expanding and enabling agile operations in companies and the extended supply chain in the new age of information.

Manufacturing Process Information

Computers have been in use in the manufacturing activity of industrial firms for a long time. They provide the control systems in the chemical industry. They provide the planning systems for most of industry. They allow for effective shop floor management. They enable machine tool automation. But computers typically have not been integrated and much time has been spent passing data and information around.

The new information systems connected to the manufacturing process will collect data and convert it to information. This information will be used to learn more about the process faster. Typically the manufacturing plant has a large number of electronic signals that can provide data. That data must be processed into a useful form so it can be used; that is, data will be collected and computation performed on it so that it is turned into information. That information will be used to direct the operation of the manufacturing activity to change and be more responsive to the customer. This, in turn, will allow for streamlining of the supply chain and a reduction in lead time. Progress and improvement in the manufacturing activity will continue at a rapid rate. Being a part of an integrated information system where the data is collected or entered once and is available to all who need it will improve the overall productivity of the management of operations. It will allow the operational information to be transactionally current, and the need for improved forecasts will be minimized. The need will be to build a capability to agilely respond to variations in real customer demands. These demands will be those that are both expected and unexpected.

The ability to improve the manufacturing activity and to make it more predictable and reliable will advance. The productivity of the plant or quality engineer will improve as more data and information are made available for the process information system. These systems will collect information and manipulate it in a way that will improve understanding of the manufacturing process. The process information systems will keep the data in the right time sequence so that trends can be developed and correlations established. This will be possible even if the data comes from analytical testing where the actual data is developed at a time later than when it was generated. Samples taken at a specific time can be matched to that time and incorporated into the data system where they belong. This allows for the samples to be correlated with the conditions that prevailed at the time of sampling.

The data collection system can bring this data together, do calculations on it, and turn it into useful information. For example, trends can be plotted, and statistical process control can determine out-of-control points. Data collection systems can also show the impact of product quality improvements that either tighten the variation or shift the center of performance. These systems that are based on data from the plant or manufacturing process can be robust systems. By using the information provided, the operational people in the plant can improve the integrity, reliability, and quality of materials and products produced, thus making the entire manufacturing process more reliable and predictable with less variation. Reliability is essential for agile supply chain operations. Understanding the manufacturing process is an essential step to mass customization.

One of the ways to make the supply chain more reliable is to provide more information for the engineers working on improving the products and processes. The following example will illustrate. After installation of an integrated process information system in a batch chemical process plant, the manufacturing engineers were able to show a significant productivity improvement. The batch units conducted different chemical reactions and operations in each of the units. The batch reactor systems produced 35 to 40 products in any of four vessels. The engineers found that they could analyze, understand, and solve 10 times the number of technical problems as was possible previously. The problem of collecting the data and formulating it so that it became information was automated and simplified. The critical path to improvement shifted from the process of gathering the data and determining the solution to that of getting the needed changes or improvements implemented. Understanding how the process was running and what to do to make it better allowed for a more disciplined and repeatable operation of the batch processes. This applied to all the products produced on the units.

An enhanced rate of learning directly affects the understanding and operation of processes. For example, it is enabling the chemical and pharmaceutical industry to shift to parametric acceptance of product quality and release for sale of materials made in these chemical processes. This involves understanding the impact of the process variables and determining if they are in statistical process control. If this is the case, then the product must be within specification. No quality testing is needed. Quality is guaranteed by the way the production plant is running. Although parametric acceptance is rapidly becoming a common practice, it does require the development of knowledge on how the process variables affect the product quality. It also requires the assurance that the process is operating within statistical control limits. This advancement of understanding and the elimination of the time required to assure quality by end product or intermediate testing are essential to having effective mass customization.

Integrated Product and Process Data

An automated information system on manufacturing processes allows for the use of statistical process control in an efficient fashion. A computer system will automatically collect the data and do the statistical computation, placing the results on appropriate control charts. Out-of-control situations will be readily identified and separated from false signals. This enables quality and plant engineers to work on finding the root causes of the quality upset and doing remedial correction to eliminate the cause. As experience is gained on the process and products, knowledge will be contained in an Integrated Product and Process Data System (IPPD). This software system keeps track of the life cycle for both the products and the process. It is the tool that captures the learning that goes on with the product and process. This learning is then available for other investigators to use as they continue to work to improve the ability to produce product in a reliable fashion.

These systems will begin collecting information about the product and process at their conceptualization and store it in a retrievable form. All events and significant information will be a part of the database. The system will provide information on the life cycle of the product and process—from birth to death and all the events in between. The IPPD system will be able to handle statistical data, the written word such as memos, videotapes, pictures, drawings, voice recordings, and any other electronic media signal. The stored information will be readily available in a very user-friendly format.

An example of this type of system was developed in the U.S. government, where an IPPD system currently is being used at Sandia National Laboratory to collect information on the techniques and art that go into designing and producing nuclear weapon systems. With the demise of the cold war, the building of these weapons is becoming a lost art. People skilled in this activity are aging and retiring, and the written word is not adequate to build effective weapons. It must be collected from the minds of the designers. The hope continues that the weapon systems will not be used, but it is a technology that needs to be captured and stored in a highly classified fashion. A robust knowledge base that allows for easy retrieval of information is a requirement. With appropriate clearances, this knowledge base is being built and used to capture all the relevant data and information. The concept of capturing life cycle knowledge does not just apply to nuclear weapons; it also has extensive potential for industry.

Industry need for an IPPD system cannot be overstated. In most manufacturing operations the same problems are solved every three to five years without the benefit of previous work or experiences. A system that allows the easy entry of solutions and logic into a database will greatly improve the effectiveness of the engineers responsible for solving a future problem. The real desire is to get to the root causes of problems and provide solutions to eliminate those causes. That is not always the case and the problem may reappear. Finding a previous solution in the database will allow the investigator to start well down the learning curve, which will make the next solution easier. This continuous capture of knowledge will result in products and processes that operate more efficiently, predictably, and reliably. The improvement in the knowledge of how products are made and how the process is run is an essential requirement in making the outcome more predictable. That predictability is required as a base for agility and mass customization. One of the unexpected events cannot be failure of the process or product to perform and produce a quality product. It is being counted on by the supply chain so that agile performance can result. These systems are not only useful in the operation and improvement of existing manufacturing facilities. Their use needs to start with the process and product development activity and the design of the first manufacturing facility or process.

The IPPD system has real value in the design phase of a new plant. Computer-aided design (CAD) drawings and information can be incorporated into the IPPD system so that the knowledge that went into the design decisions are part of the initial information on the new plant. Capturing the logic that has gone into the selection of a piece of equipment for the plant allows future people who deal with the maintenance of that equipment to understand more about why it was selected. For example, a design engineer in the chemical industry was questioned on his logic for the particular choice of a process compressor. He gave a very detailed and understandable explanation. The IPPD system would enable that explanation to be put in a memo that is retrievable and forms the starting point for exploring the performance of the compressor in the later phase of its life. The same capture of information on decision applies to the way the process will be controlled. The logic for the way a control system is set up is important in understanding how it was intended to work. Process control systems differ by different industries but they exist in all. The building of the process control system can become very personal to the designer. That individual uniqueness needs to be the understanding that gets passed on to future engineers on the process.

Process Control

Manufacturing processes have process control systems to ensure the desired outcome. They also ensure that the manufacturing process operates safely. The process control system is made up of a data and information system, and a process control system. The data and information system installed on operating manufacturing process or machines is a critical part of the management system. It is not only used to gain understanding of the process but also is critical in making sure the process or machine is working the way it is intended.

In the chemical industry the foundation for the process control system is called a Process Information, or PI, system. It collects information on the process's performance. The process itself is controlled by individual loop controllers, which control things like temperature, pressure, or flow. These individual controllers are usually built into a control system computer. Historically, they were stand-alone units that worked independently on only one of the control loops. A chemical process might have a thousand loops. Today the PI system collects data from the process and turns it into information about the process. It then analyzes the information and determines what course of action to take. Working with mathematical models that optimize the material and energy balance of the process, a PI system has the ability to predict future events and determine the desired changes that will optimize the process performance. It understands the restraints of the process and manages within them. The model has output that resets the loop controllers and thus changes the way the process is being run. This model-based control can have a dynamic nature across the supply chain. If a feed composition changes, the model can predict, based on time constants, when set points need to change through the entire length of the process train and the many steps that are involved. This ability to control the operation of the chemical process provides an ability to reliably predict and optimize the process performance. This is critical to having a process that can give an agile response from a single process train. Balancing the product distribution to unexpected variations in demand is an agile response required across the chemical and petroleum industries.

Other industries have the same type of overall manufacturing train management systems. They are a combination of controls on individual machines. These controls drive the machines to make the desired and sometimes varied products. They also provide a knowledge of the material movement progress and integrate the train. Work cells or manufacturing machines can then be coordinated in an integrated fashion. The computer can replace the Toyota *kanban* systems. When the supply chain is extensive, computers coordinate the steps in the assembly process and also the activities of the supplier network. This is done in a just-in-time fashion with a minimum of inventory. The computer's capability to do this effectively enables the supply chain to respond agilely to unexpected change.

This control system does not stand alone. It is a part of the whole supply chain and thus provides information to and takes information from the other parts of the supply chain. Having each step of the process fit into the whole is critical to the effective operation of the supply chain. The status of the equipment or process units in the supply chain is important to the people who translate demand into directions for the manufacturing activity. The status of material or product in the supply chain is critical to the operation of the transportation system that serves the factory. The determination of when shipments will be made and when they will be received hinges on accurate knowledge of what is happening in the manufacturing systems. Thus, all elements must work together in an integrated fashion to be able to respond to unexpected change in an agile fashion.

Integrated Enterprise

The computer is the tool of choice when developing an integrated supply chain. It provides for optimized instruction for the running of the manufacturing part of the enterprise. It also works to integrate the customer and supplier base with the supply chain. In many companies, the suppliers have access to what is happening in their customers' manufacturing units and the scheduling activity. This relationship enables the entire supply chain to operate as an integrated unit. The knowledge of what is going on and when something is needed is critical to making the whole run effectively. It is also critical when the schedule changes due to an unexpected but high-priority request. The entire supply chain can see the request and adjust its activity so the change is handled in an agile but routine fashion. The interactive communication that occurs based on the proper information being available allows the entire supply chain to respond with agility. The computer puts the discipline into the system that makes this happen. This integration is being provided by enterprise resource planning (ERP) software. These systems are replacing the functionally developed software of the past.

Today software exists that allows for large integrated systems that span the entire supply chain and integrate the running of the enterprise. For example, a software product like SAP's R-3 provides a backbone for the integration of the various activities of an enterprise. This backbone integrates various modules of activity so that operations can be looked at as a whole. Data is entered only once and then is usable by all modules or users. The integration system provides for all people who need information to have the same information. It does the job of passing information from one decision activity to another and automates the management of information and decision making. It makes decision making more routine and disciplined. The systems are usually transactionally current, allowing the latest information and data to be used by the entire supply chain. The business transactions that are supported are many and comprehensive. They typically take the shape of modules that represent all the activities of the firm, which include everything from personnel to maintenance. Modules are designed to be compatible with the integration software. In many cases they are supplied by other companies that are sponsored by SAP. It truly is a robust system with software capability in a constant state of improvement.

The modules need not be software that comes from the integration system supplier, but they must be compatible. This has led many software houses to perfect a particular module to fit the integration software. Purchasers of systems and modules will be looking for the "best in class" for both. They will want to have the best capability possible as long as it comes from a software house that will be around to update, improve, and service the software. Figure 5.1 is a list of modules that are in a typical system.

Additionally, the system will interface with a Process Information (PI) system and an Integrated Product and Process Data (IPPD) system. The complete system is necessary for a firm's operation. If the system is installed and operated properly, a firm's effectiveness and productivity will be improved. It will allow a firm to focus on and improve lead time along the supply chain, because the speed of the supply chain will be limited not by the speed of developing and managing information and using it in decision making but by the speed of the capability to produce and deliver. The many modules work together to create this overall improvement in efficiency.

As an example of what the modules might contain, the quality management module monitors, enters, and manages work processes relevant to quality along the entire supply chain. It coordinates inspection and testing processes and the approval of lots. It monitors the corrective measures processes to determine when products are ready for release to the market or customers. It also monitors and provides information on status of laboratory activities and their qualification. It does all the quality information system management. In doing this, the quality management module integrates quality inspection into production. It manages the quality certification and notification systems,

> **Figure 5.1   Modules for an Integrated Operational System**
>
> Sales and Distribution
> Financial Accounting
> Controlling/Cost Accounting
> Asset Management
> Project Systems
> Human Resources
> Plant Maintenance
> Quality Management
> Materials Management
> Production Planning
> Workflow Management
> Warehouse Management
> Environmental, Health, and Safety Management
> Treasury and Investment Management

records the defects, does statistical trending of data, and uses statistical process control throughout the system to ensure that reliable information with significant variations are highlighted. The quality management module is a robust system that will be developed further as software is improved to automate an activity that has historically spent a lot of time manipulating and transposing data. The result will be a reduction of supply chain lead time and an increase in the quality of product that is shipped to the customers.

In the management of financial activity, the controlling/cost accounting module is usually a complete system that harmonizes planning and control instruments and enterprisewide controlling systems. This module documents the entire value flow of the organization, its cost structure, and the factors that influence these. Functional features of the controlling system are things like product cost control, profit analysis, activity-based costing, project- or contract-based billing, cost center accounting, and budgeting and management consolidation. These have historically been processed monthly, but the capability exists to operate this system in a transactionally current fashion.

Production planning is another one of the modules and provides for many types of manufacturing activity in a comprehensive fashion. It includes production of repetitive products, make-to-order, assemble-to-order, make-to-inventory, and process-oriented manufacturing. It uses extended MRP II, electronic kanban, control systems for process and machine, CAD, and PDM. It includes production scheduling, material availability, and equipment availability and interfaces with the plant maintenance module as well as the material management module. It also does capacity planning and optimization. This is another robust module that will get better as improvements occur.

The other modules are equally robust and are important in how they will automate the information system in an enterprise. Step change improvement will enter a continuous improvement activity as the systems are installed. The modules will do much to automate, integrate, and make the systems work more efficiently. It is anticipated that jobs will be eliminated and people will need retraining when these systems become operational. It is also anticipated that the supply chain will have a reduced investment in inventory and a higher asset utilization because of the optimization potential of the system; that is, the enterprise will improve its leanness. The question might be "What will these types of systems, as good as they are, do to agility and mass customization?" Although they are not being built for mass customization, the method that they use will lend itself to agility. The module approach of a single data bank for a supply chain should enable the user to adapt to opportunity by the type of module installed and the way it is used. As industry gains more experience with an integrated information system, the system will enable, not hinder, agility and mass customization trends.

SAP is not the only firm supplying this software but it does have the largest market share. The concept of an integrated supply chain is critical for the benefits that are possible. The legacy systems that have been used will not have the speed and completeness to be able to compete with the integrated systems that span the entire extended supply chains.

Scientific Computing

One of the areas that is being significantly influenced by the revolution in computer capability is scientific computing, a very complex field with many important variables. Historically, the scientist has limited the number of variables that one studies and in many cases discovered knowledge that was important but did not fit what was happening in the real world. With the fast and large computing systems that exist today, many more variables can be studied in conjunction with the primary ones. The science is getting closer and closer to reality. It is anticipated that future computational capability will allow even more robust modeling and understanding to occur. Numerical intensive computation is no longer limited by the computer or its cost but by computational techniques and the process of defining the problem so it can be understood. This advance in the speed of computation has challenged the scientist and computer people to develop the problems and the solution techniques that address scientific challenges that have not been solvable in the past.

In the chemical industry a workshop was held at the University of Maryland in October 1996 by the Department of Energy (DOE) on one topic, Computational Fluid Dynamics (CFD). The DOE Office of Industrial Technology (OIT) was attempting to integrate the needs of the chemical industry with the capabilities that have been developed in the National Laboratories. A team of academics, industrial scientists, and National Laboratory technologists worked for two days to define the most important focus for combined work. The National Laboratories are spending $100 million in the area of CFD for a broad range of systems that span all scientific activity and all industries. This workshop was aimed at focusing some of those resources at the chemical industry. OIT put up some money to support the programs inside the government laboratories while industry would provide money as work in kind from its programs.

The two areas that were selected by the team were mixing in the chemical industry and particle dynamics in things like fluid bed reactors. These are major challenge areas for the industry, but by working together, progress and understanding will occur. The results will make the design of new systems more understandable and the operation of existing chemical processes more effective. The robustness of computer systems will make this technology development a reality.

Other areas where the computer has had an impact are robotics, machine tool control and management, molecular modeling, and finite element analysis. This is not an exhaustive list but illustrates the nature of the computer revolution in the scientific computing area. The scientific community leads the way in defining high-level modeling and computation that will eventually work into the software of the commercial world. The commercial world should anticipate that each of the places where computers are used will become increasingly capable with more powerful knowledge-based computer systems. How long will it be before the capability will exist that allows the anticipation of unexpected events? This is required before one can respond in an agile fashion. It may be that processing demand information or trends will allow for more reliable forecasts of future demand. Assessing large fields of information and then using this assessment to develop very robust models will enable the operational people in a supply chain to predict what will happen. With a better view of what will happen, the agile capability can be prepared to respond. It is easy to see that the more typical commercial activity will be capable of responding more quickly to the unexpected event. The information system will provide the knowledge and direction that allows the execution of an effective response.

The application of this scientific computing power to the supply chain and commercial activity will significantly change the breadth and scope of activities that can be managed effectively. Routine decisions will be done automatically in a more capable fashion than if done manually. Likewise, more complex relationships will be more easily managed. As improvements in the scientific computing field emerge, the operational integrated management system will become more agile and able to easily adapt to different managerial situations. The various modules will become more effective, and things as simple as budgeting or accounting will be changed significantly in line with the agile enterprise. Just what all the changes will be is yet to be determined, but the power of the computer will allow almost anything to happen. The challenge may be to develop the people and skills to go along with what is possible with the computer and the expected advances in software.

Summary of Integrated Information Systems

This chapter has talked about the use of computers in the commercial and scientific worlds. Industrial agility will benefit from the computer revolution. For example, the practical concept of mass customization at Anderson Windows, described earlier, shows the role that the computer can play to integrate the supply chain from the customer well into the manufacturing activity. Without the computer simplification and integration, it would have been very difficult to develop a custom window capability.

A significant contribution of the computer has been the enhancement of the reliability to produce product. Making it right the first time is assisted by learning how to produce and maintaining that learning for others. Learning curves truly build on previous experiences. This knowledge results in the development of parametric-based process control concepts and process control that ensures the product meets expectation. The computer plays the key role to make sure the machines or process run the same each time. They even do self-diagnosis for wear or drift of control instrumentation. They provide for automatic calibration and validation of testing.

Integration of the supply chain provides the discipline needed to ensure that the desired results are attained. This reproducibility and reliability relative to the operation allow for agile capability to be effectively used to serve each customer as desired. Integration is a key for properly determining the demand and then translating that into a production schedule for production and delivery of the product. Make-to-order with custom products is a reality and takes an agile interface with the customer. It must be supported equally by an agile capability to produce and deliver.

The next chapter will discuss obtaining the capability to serve the customer agilely. Developing a capability to respond agilely is something that demands considerable assessment and discipline. Responding quickly and effectively to unexpected change requires forethought and planning. It also requires the capability to meet the range of possible change in an anticipated fashion.

CHAPTER 6
Changing Capability

Building agility into the supply chain does not occur without significant planning and effort. Chapter 2 described the process of managing change in an enterprise. Motorola and Anderson were highlighted as examples of companies that excelled at managing change and building agility. This chapter deals with providing capability to the processes, products, markets, and supply chains that will make companies more responsive to unexpected change.

Agility is one of the tools that supports the way a firm decides to do business within a specific supply chain. Developing a vision of how commerce will be executed is a very critical step in developing capability. If the vision consists of doing business the way it has been done in the past, then new capability need not be considered. If the desire is to change business practices, then specific initiatives are essential to achieving that vision. These change initiatives will probably deal with the speed at which the supply chain operates or the responsiveness to the customer. Speed is measured in terms of lead time. Responsiveness to the customer is measured both by how quickly and how well the product meets the customer's needs. Change initiatives may address both these areas. They are aimed at increasing the competitiveness of the supply chain. The commercial or supply chain management team starts the process by assessing the existing supply chain and determining whether change is warranted. The team must also determine what change is desired and what benefits the change will provide from today and into the future. These two steps start the process of developing the agile or custom manufacturing supply chain capabilities.

Today and Tomorrow's Capability

Building a new supply chain capability to serve customers starts with assessments of how business is being done today. This is followed by determining the way it will be done in the future. Then the challenge is to develop the capability so that the business can be run in the fashion desired.

To do the assessment, a company must understand its business as it exists today. This assessment can expose undesirable situations such as poor profit contribution, declining sales or market share, poor utilization of assets or capital investments, too much investment in the supply chain cash cycle, product offering that does not meet the needs of the customer, and so on. These are the symptoms that lead a company to believe that something needs to change. There is one other motivation for change. Business is good and the company is making money, but if it could improve the supply chain, then it would grow faster and make more money. The first need for change is a recovery situation, whereas the second is a preventive or competitive enhancement situation. In the preventive situation, change is more difficult. In the recovery situation, the threat of disaster hangs over the supply chain and that tends to make change easier to accept and accomplish. Without the threat, a powerful vision with defined rewards is needed to initiate change. In the parlance of Joel Barker and other experts on change, an understood "perceived advantage" is necessary. When things are going well, it is hard to get consensus on the advantage of the change. Establishing a vision of a different way to do business in the future is helped, however, by the shadow of impending disaster looming over the company.

The vision of how a firm or supply chain will do business in the future to improve its effectiveness requires the definition of several elements that pertain to the commercial situation. An example of an element might be: "Do we want to supply the customer using make-to-order or make-to-inventory?" The marketplace is part of this determination. Is the demand for a broad product line with custom products a significant part of the market? If this is the case, then a capability to supply quickly in a make-to-order fashion might be important and a key element of success. The capability would need to be developed.

Make-to-order can be done in a number of ways. For example, a standard product can be made after the receipt of the order. Make-to-order can also be the production of a custom product tailored for the customer. In either case the objective is to supply the customer what is wanted in the quickest time possible. Just how quick is quick enough is determined by the specific market. In some cases, the production time might be a year; in others, production needs to happen in minutes. For example, the process of producing a custom-designed machine that is made to fit the specification developed between the customer and the machine manufacturer might take a year or more before the machine is up and running. But a computer-generated design on a custom-made T-shirt may be applied in minutes. Thus the situation will dictate the production and delivery capability that needs to be developed. The trend will be for the response time to make-to-order to get shorter and the customer to require more service upon the receipt. To be effective in satisfying the customer, the manufacturing and delivery part of the supply chain must be assessed to determine where product will take on its identity.

The supply chain must be studied and the right place determined where the product can be converted from a generic intermediate to a product with its identity established. This varies by supply chain and product technology and is an important consideration. The closer to the customer that the generic intermediate can be maintained, the quicker the response can be and the more effective the supply chain will be. In a make-to-order or custom business where the product is standard and the final production occurs only after receipt of the order, then it is desirable that the last step be one that converts the generic intermediates or materials to the product quickly. An example of this might be the providing of custom colors from a paint store to the retail market. A generic paint color like white is inventoried along with an assortment of pigments. After the customer selects the desired color, the gallon can is opened and an automated machine puts in and mixes the right pigments. The customer leaves with the desired color that came from a standard capability. The capability was developed to minimize inventory and to solve shelf life problems. It also gives the customer a wide selection of colors from a finite number of pigments. Postponement of the determination of the final identity of the product to the last minute makes the supply chain more effective by offering the customer a lot of variety with a minimum of inventory.

To gain even more variety in the offering to the customer, it may be desirable to move back the point where the identity of the product is determined in the supply chain. This requires that the steps between the generic intermediates and the customer be streamlined, sped up, and made more reliable for the supply chain to work. An example of this would be a company making rubber compound that is sold to people who make it into cured rubber parts. The rubber formulation determines the properties of the final part. The process to provide the compounds starts with a generic polymer being processed in a large batch mixer with other components of the formulation until it is completely mixed. This is then sized properly and placed on a two-roll mill where the catalyst and pigments are added. Before packaging, the compound is strained to eliminate grains and grit and shaped for the specific customer. This process provides a very large variety to the customer but has run with a great amount of inventory in a not-so-streamlined fashion. By having the customer's order determine what will go into the large mixer where the formulation is developed and then having continuous flow processing of the customer's entire order to completion, it is possible to eliminate a significant amount of inventory and still maintain variety and customer response. Determining how to make the last step in the process a continuous flow one was the critical issue. Streamlining allowed the company to maintain what the customer desired and also improved the cash productivity in the supply chain.

The change was made in small steps or continuous improvement. The capability to deliver this type of custom or make-to-order product needed to be developed and proven. It enabled the rubber compound provider to respond to the daily demands of the customer with a three-day delivery from receipt of the order.

Another aspect of the decision to change capability is to understand the daily demand of the customer. This is not the demand from the warehouse but reflects the customer order patterns. It is important to get today's customer order patterns and give them a critical assessment. In many cases the patterns are influenced by things that companies or the distributors do. The real desire is to understand the use patterns of the customer and try to level out the ups and downs of the supply chain.

The use pattern for something like washing detergent, for example, is probably fairly constant. Historically the purchase patterns for the product depended on the deal that was available in the supermarket. In the case of Procter & Gamble (P&G), the company was selling almost always from deals or promotion specials. The customer demand was managed, and customers quickly learned that the time to buy was during the specials. P&G was using the marketing strategy that specials built demand for its products ahead of the competitors. The supply chain of P&G needed to respond to this type of demand. The assessment of the cost of doing business this way resulted in a change in the way business was done. It was driven by optimization of cost and the need for a large capability to handle the significant surges in demand that resulted from specials. The capability to handle the surges was many times greater than what was needed to handle material flow consistent with the real level of consumption. This story and the resulting change to a leveled-out and nonspecial-based demand was described in a *Fortune* article titled "The Dumbest Marketing Ploy." Procter & Gamble shifted the way it did business and really improved the supply chain effectiveness. The system is much reduced in size and does not need to rise to artificially stimulate demand. Profitability improved for P&G, and its competitors all followed suit. The consumer benefited from overall lower prices for excellent products.

Besides promotions, the terms of doing business will drive an artificial demand. For example, terms that promote purchases of large lots of material can distort the real customer demand. Freight terms and who pays them cause another distortion. The demand must reflect as closely as possible what the customer will be buying and using, because it will determine the basis for changes in capability along the supply chain. Historic demand without understanding what the patterns of product use are can lead to the wrong conclusions and thus the wrong approach to the market. The stimulus for doing something to promote growth or capture customer loyalty must be understood and the merits analyzed. The trend is to do business the old-fashioned way: Deal with the customers in terms that they prefer—high-quality product that meets all needs, is delivered promptly, and costs what customers would expect to pay. Many firms have made a business from this approach. Improvements can come, however, from building the capability to do this more quickly and with the variety that the customer desires. That capability must have the agility to respond to the customer usage or demand.

The Customer Order

A critical decision in developing capability is determining how the demand will be established. If it is make-to-order, then a certain type of capability will need to be developed. If it is make-to-stock, then a different type of capability along the supply chain will result. With the approach determined, various changes in capability can occur. The capability to take and process orders in an effective and efficient fashion must either exist or be developed. When products come from the warehouse, there is not the same need to process the order and drive the production process. The customer order will drive the pick-and-ship process at the warehouse. The production activity will be driven by the replenishment order when safety stocks are reached. This does not mean that the process of picking and shipping is not vital for the success of the firm.

Companies like Land's End or Jackson and Perkins must have systems that respond to the customer in an appropriate fashion. Land's

End must ship to the customer as quickly as possible. Quick customer service is a trademark. Jackson and Perkins needs to let customers know just when nursery items will arrive at the customers' location. This date varies with parts of the country and the seasonal changes. It's a challenge to get roses to Michigan just at the appropriate time to plant as well as to Tennessee when planting is appropriate there. The order-driven pick-and-ship process must be designed to fit the need.

A different system is required when make-to-order is the type of demand. The order system drives both the pick of raw materials, the produce or make, as well as the ship processes. This requires a system that immediately incorporates the order information into the production scheduling activity. If delivery is going to occur in a timely fashion in a make-to-order fashion, then time is needed to produce. The shipping step must occur immediately when the goods are ready. A make-to-order system can work for both standard products and custom products, although different capability needs to be developed in the production capability to support each type.

Customization

Custom products can be collaborations between the customer and supplier. Four types of customization were identified in an article by Gilmore and Pine in the January-February 1997 issue of the *Harvard Business Review*. For example, Eye Tailor collaborates with the customer in providing eyeglasses. The company works with the customer collecting information and then fits glasses to a digital image of the customer's face. Adjustment of sizes and shapes occurs until the customer is satisfied. Finally, the lenses are ground and installed in the frames. The customer leaves confident that a new image has been created—all in less than an hour.

A second type of custom product is adapted by the customer for a unique use. These products can range from lighting systems to mattresses. The lighting systems are filled with options that let the individual establish the mood that is desired for a specific activity. The mattress might have air chambers that can be adjusted to control firmness. Another example would be sound systems where the listener is allowed to select the pieces to listen to and also can provide adjustments to volume, balance, and tone.

The third type of custom product is the cosmetic approach where the presentation of the product is adapted to the customer or market. Packaging a sealant to the marine, home maintenance, industrial, and electronic market but filling the package with the same material with different graphics is an example. Selling a T-shirt with a stylized symbol on it that fits the customer's need is another form of cosmetic customization. Hertz does this with its Gold Service, where each customer's preferences are preestablished and then delivered in what appears to the customer as a very personal way. Hertz found that this method cost less than the normal way of providing the service where the needs were determined over and over again at the time of rental.

The fourth type of customization is transparent. In this type, the supplier of the product observes the customer's use of the product and then adapts its response to the needs that result. A hotel chain that determines what the customer preferences are and then provides a room that meets those requirements in a routine fashion is an example of transparent customization. The number of personal preferences that the hotel responds to is determined by the demand of the regular customers. The demand may be as simple as smoking or nonsmoking rooms or the right type of pillow. The customer is not aware that these amenities are being provided. Vendor-managed inventories tend to be of this type, where the customer will take for granted that the material or part will be available when it is needed. The vendor monitors use patterns and preferences and develops an effective system to ensure availability.

Customization is a key part of the capability that needs to be developed for the specific supply chains. This trend will undoubtedly expand in the year ahead and may be the key element of both retail and industrial activity in the decade ahead. Becoming effective at customization will require understanding the four types and adapting them to the situation for a particular supply chain. In most cases agility will be required for both the custom capability and the relationships that are involved. The relationships must be built for the circumstance. A vendor-managed inventory relationship is much different than one where a vendor responds to an order. There is trust in both relationships, but the one where vendors manage the inventory requires more.

Satisfying the Order

The customer order pattern, the make-to-order approach, and the degree of customization determine the capability that must be utilized or developed. They are key parts of how business will be done and drive the amount of equipment and machines that are needed. The peak demand in any one day must be determined and the order patterns for the family of products must be looked at statistically. The supply chain team needs to determine what portion of the orders will be satisfied in the desired time. The delivery to customer request date provides a key measurement of this performance. Satisfactory service is usually provided when the measurement is from 95 to 98 or 99 percent. Studies have shown that customer service above 92 or 93 percent, however, is not detectable by the customers. The investment to cover all orders in the desired time will be high and the team needs to make a trade-off of cost versus service. These trade-offs must be understood and a disciplined approach developed so they are made in a consistent and uniform fashion. Parts of the team will want to fill all orders on time. Others will want to fill most orders on time. Still others like the concept of make-to-inventory and the long production run of a single product. Putting numbers and measurements on this activity will facilitate reviewing the performance and determining whether it has been adequate or more capability needs to be developed.

With agreement on the expected level of demand, the desired performance, the concept of make-to-order or make-to-inventory, and the degree of customization, the team can develop the capability to satisfy the customer. It can determine the elements that will be critical for the supply chain to perform. Specific attention needs to be focused on equipment or machine capability, information system requirements, how the system will operate and respond to the order, and how the product is shipped to the customer. These choices in capability development raise alternatives that have economic implications. For example, a direct investment can be made that provides the capability to cover the demand. This investment may be different than what is in place for make-to-inventory. The process or machine will require changeover with every order. The various components must be able to be delivered as demand dictates and consistent with what will be produced. The entire raw material or component delivery system must be integrated with the orders or demand, production scheduling, and production operations systems.

Another feature of capability is that the product will need to be made right the first time. Reliable product and process technology must exist. Defects must be at a very reduced level and rework cannot be frequent. The product must have the desired quality from the capable process and machines. Motorola's six sigma approach is an essential for industry. It is a requirement of the future. It does require a special mentality that drives the change to improve capability. To operate in a make-to-order fashion with quick delivery, this reliability to produce must exist.

Besides installing the process or machine to meet the peak customer demands, there are some other choices depending on the circumstance of the production technology. Since demand for various different product families will differ, the sharing of the production capability could occur. It involves determining whether the process or machines can be interchanged to make both product lines. This might require some modification of the equipment so that this interchange can occur. If this is possible, then the two product line management teams can analyze the demands from the customers together and determine if synergy will allow one process to act as backup for the other. A third unit could even be involved. If only one of the product lines will be make-to-order and the others are make-to-inventory, then even more flexibility exists for this approach.

This concept was used in the chemical industry where the equipment was batch kettle reactors. These were not simple units. They were tailored to the product lines that would be produced in them. Some had distillation or stripping capability; others had filtration systems included. Raw materials were added differently to the equipment. Each was a custom-tailored unit made to do a specific task and make a given product. Because the product lines were different and each had the peaks and valleys of demand, the equipment utilization was 45 percent. Thus, there seemed to be a clear possibility of sharing the equipment among the various product lines. The team from the supply chain and the engineering departments looked at what capability would need to be changed so that all products could be made interchangeably and reliably on the three kettle reactors. The necessary modifications and standardization were feasible. After changing the capability, the production schedules had an increased degree of freedom, and more overall capacity was available for all three product lines. In this case, the shift to make-to-order was accomplished for two of the three product lines. That shift caused an acceleration in product growth, and the asset utilization climbed until it was practical to include another kettle reactor in the system. Utilization is now at 75 percent and growing, up from the preciously mentioned 45 percent. This change had required a number of things to happen. First, product technology needed to be reliable, and some work was needed to eliminate the root cause of defects. Second, demand needed to be understood for all the product lines, and some artificial stimuli had to be removed so that real customer needs were reflected. Finally, the raw material supply capability to the three units was integrated with the customer demand, and material handling technology was improved.

These changes to the capability to make are not all that is required to have a make-to-order supply chain. The product needs to move to the shipping dock and then on to the customers. This requires a lot of attention. In the example of the chemical plant's batch kettle reactors, a series of loading spots had be added for more direct-ship trucking. The packaged material came to the dock on pallets and those pallets were aimed at specific industrial customers. On the dock they were placed in the right truck going in the right direction. This was done in the right order so that first in was the last thing to be delivered. The trucks went to the customers, and all the goods were delivered within the next 24 hours.

The make-to-order concept that started as a vision changed the entire supply chain in an agile way so that the customer could be served better. Customer service is one of the important benefits of this type of change. It is hoped to result in profitable growth and building customer loyalty. Another benefit is the improvement in asset utilization. The chemical industry example described a movement to a higher overall utilization of process equipment and the material movement capability. This gives a higher return on investment. Another benefit is that the system eliminates finished-good inventory and, if done effectively, in-process inventory. The improvement in process reliability can reduce rework inventory toward zero. In most cases of change, the new capability will increase customer satisfaction and make the financial performance of the supply chain much more acceptable. The benefits are waiting to be captured and the use of agility thinking can be an important part in the changing of the supply chain.

The Three Flows

The preceding discussion dealt with a vision and a shift from make-to-inventory to make-to-order. Many more elements of the vision, however, must be looked at as having benefit. One of those might be electronic commerce. If the need of the supply chain was to shorten the cash cycle and get payment earlier, then the team might consider electronic commerce. This is not the only reason for this consideration, but it is an important one.

The vision the team develops might go like this. The movement of material and products in the supply chain will be monitored and managed using a bar code system with particular tags for each component and for each product. The tags can be assembled into an order and a consolidated tag can be assigned. The shipment of product leaves the dock of the producer and arrives at the dock of the customer. It is scanned. With the scanning the trucker can then go on to the next stop on the delivery route. The tracking information is then fed to the computer of both the receiving company and the supplier. A command is then issued in the computer to transfer funds electronically according to the prenegotiated terms. Those terms could be immediate payment or payment in 30 or 60 days. The price may vary depending on when the payment is made. The customer will want the most generous terms possible while the supplier will want immediate payment in full. The driver in both cases is the cash cycle—the time between when materials are paid for and when cash is received from the goods produced. The shorter the cycle, the less investment that a producer will have tied up and thus the higher the return. Managing this in an agile fashion with each customer treated appropriately can have significant benefit. The way to achieve maximum benefit would be to shorten the material flow, the information flow, and the cash flow. Agile capability is required for each of those three flows. They must be able to respond to unexpected change.

Material Flow

The three flows each have unique unexpected changes, and capability must be built into the systems to handle these events. For example, a trucker may spend time off the road with a friend and not get the material delivered according to the schedule and the customer expectations. This happened in Europe with a shipment from Belgium to the United Kingdom, in a supply chain where critical material was routinely delivered in a just-in-time fashion to a soap producer. Tracking the shipment down became a real chore with the trucker out of circulation with his friend. An interruption in supply caused a down time in the customer's process. Capability needs to be able to react to such unexpected events when priorities change because of unique opportunities. In this case, the capability had not been developed and it set people looking for how it would be handled. A contingency plan was developed, and the system to instruct and track the trucker was improved. Other nonexpected events could cause a material flow interruption. For example, weather could affect an ability to get goods to the customer. These events must be anticipated so that an agile response can be developed. The material flow capability illustrated here is a key part of the "Three Flows." This is shown in Figure 6.1.

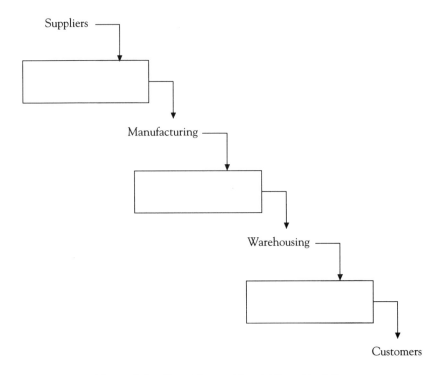

Figure 6.1 Three Flows—Material Flow Capability

INFORMATION FLOW

Material flow has its unique problems as just indicated. Capability must also be developed to anticipate an unexpected change in the flow of information. Simple things like inaccuracy of bar code reading or labeling can throw a computer system into disarray. This type of failure has resulted in systems being developed that allow the printing of bar code labels at the point of packaging or origin. The ability to ensure the proper label and make the people on the shop floor responsible required this capability to be developed. With systems that have thousands of labels preprinted, the error most frequently encountered is selecting the wrong one. This error can be eliminated or reduced by using computer information systems that integrate the supply chain and the computer that knows just what is being made. The computer will only allow the right label to be made. The check on the system is

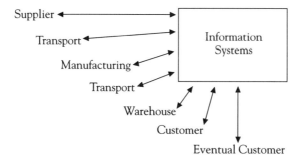

Figure 6.2 Three Flows—Information Flow Capability

the operator on the shop floor. This is one of the things that will improve the integrity of the information system. That integrity is needed for make-to-order, make-to-inventory, or custom products. Agility does not mean that response is freewheeling or that reliability must not exist. In fact, it demands that we know more about how to do business and that actions are reliable and predictable. This is shown in Figure 6.2.

Cash Flow

The flow of cash also requires a high degree of reliability and predictability if the system is going to work properly. The bar code input must be right. It drives the actions of the customer. One of those actions is the release of payment. The system must have the proper checks and balances to ensure that the cash payment can be made. Integration of the information system with operations is required. The product must be anticipated, and it must be required in the production schedule. Evidence that it has been used must also come from the electronic run sheets. The transfer itself must be supported with cash, and confirmation must exist that it has been made. Electronic commerce is not as simple as plugging two banks together. If it is to work effectively, the cash or money flow of the supply chain capability must also be developed in the same way as the other flows. This is described in Figure 6.3.

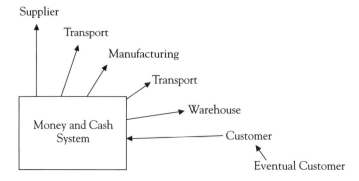

Figure 6.3 Three Flows—Money Flow Capability

In the preceding example one can note that the electronic system is an integrated extended supply chain activity. It integrates across the internal supply chain and to suppliers and customers. This implies that capability exists at all three firms and that the capability is compatible. Thus, a partnership and cooperation exist between the firms along the supply chain. It also means that parallel systems are probably necessary because of the variation in capability among the customers and suppliers. Progress to the systems described does not occur instantaneously but requires a vision and persistent sustainable actions toward the goals.

Product and Service Quality

Another element of a vision might relate to the highest quality product being in the hands of the customer. This could be a competitive advantage. There are two contrasting situations in supply to the consumer market. In one case, freshness is what sells the product, and the grocery supply system provides for that. Things like vegetables, meats, fish, bread, dairy products, and potato chips are all examples of products that are better when fresh. They have dates when the product is out of shelf life and should not be sold. Even beer has taken on the campaign that freshness is better and certain companies label the product with the date it is produced. The capability has developed and improved over time to ensure the customer gets the freshest possible product.

In another case, freshness has not been a paramount issue in the consumer market. The following discussion will illustrate, however, that grocery store chocolate candy is an item that could benefit from a change in the supply chain capability to make a better product available to the consumer. Fresh candy is indeed better than that which has been around for extended periods. The people in candy factories carry the lot codes and dates so that they make sure they snack on the freshest candy made. Nuts in candy are a specific challenge. They age faster than chocolate and change taste more rapidly. The candy industry has done a lot to try and improve the time that candy retains the best taste. For example, companies that produce the more expensive candies provide refrigeration to ensure freshness. Mass-produced candies, however, are faced with a cyclical demand with major consumption times at Halloween, Easter, and Christmas. The big candy companies look at their supply chain as one where freshness and the super taste that goes with it are not important to the customer. They feel that the customer will snack on almost anything as long as it is chocolate. Thus, the big candy companies run the candy and packaging machines at capacity all the time. They have an optimized lean operation with large warehouses and a supply chain full of candy just before the demand surge around Halloween. They produce the candy from April to October, so some of it is six months old when delivered to the grocery shelf. It has gone from a delicious tasting morsel when fresh to one that tastes sweet but is not very satisfying. The consumer is deprived of some satisfaction.

This segment of the candy industry has a choice that would probably cost less in the long run and provide consumers with a more pleasant experience when they stop for a sweet treat. The industry could look at its supply chain and apply the concept of potato chips; that is, expiration dates could be used that are consistent with acceptable shelf life. If the candy does not sell, it could be brought back in-house for disposal. This new vision for the industry would require building the capability to accomplish the necessary change. Determination of what the acceptable age of the candy is to provide the right taste is information that the companies have. This would become the basis for the development of other capability.

Such a change would probably be the end of the large warehouses holding Halloween candy. The reduced investment in candy inventory would require an increased investment in packaging and production equipment that can meet the demand of the consumer. The industry is very automated and more of this candy making and packaging equipment would need to be installed. Since it would need to run at various speeds throughout the year, a larger workforce would need to be developed to make more candy in September and October than has been done before. The delivery and return system also would need to be different when packages are labeled with shelf life dates. Consumers will understand that fresher is really good, and they will scan the shelves for the freshest candy. It surely is a shift that would make a good product even better, but would require changing the many paradigms that prevent this method within today's candy industry culture. Competition is not keen enough to use this tool to challenge the two companies that dominate the industry. The supply chain can be developed to respond agilely to the customer demands and needs while still maintaining the profitability of the consumer segment. A shift in competitive situation would indeed be unexpected. If it occurred and the consumer relished the change, one company would be unprepared for the surge in growth and another would be wondering what happened to its sales. Candy lovers would benefit.

Summary of Changing Capability

Building capability so that supply chains and enterprises can attain a vision to make a company more competitive requires the use of agility thinking. To have a responsive supply chain will require the development of a number of capabilities in areas that on the surface don't seem to support agility; that is, in the material, product, and information areas. Improved supply chains do not become agile without building an agile capability. Agile capability does not occur overnight; it needs to be done in a step-by-step process.

Nevertheless, the process of changing the way a supply chain conducts business is an important part of the shift that is occurring in enterprises today. The advent of robust computers and the information systems that they enable has stretched the envelope of potential options in doing business. The changes are focused on customer satisfaction improvement and increasing the speed of information, cash, and material flow in the supply chain. The process starts with the assessment of today and then determines a vision for the future. The vision must address each of the important or key elements that determine the way business is done. Some elements will need to change, whereas others will support plans for the future. With the vision established, the capability to do things differently must be developed. Although the execution of the plan to change must be looked upon in its entirety, change actions must be staged. Remember that the change to doing business differently is a journey and not something that will occur easily. Focus must be not only on the technology of change but also on the human organizational considerations.

This chapter has dealt with the process of building the capability for changing the way a supply chain or firm will do business. The next chapter deals with how a changed supply chain or firm can operate. It brings forth the concept that agility in the supply chain requires automation. The four areas of automation will be discussed. These areas of automation will be the machine or process, the movement of material, the system of information, and the movement of cash.

CHAPTER 7
Automation

Automation has been occurring throughout modern industrialization. It has influenced both mass and lean manufacturing. From the time of Henry Ford and his assembly lines, automation has been increasing in the workplace. It not only was introduced in the movement of parts or assemblies but also became a part of the machines that produce components, parts, and product. Robots were introduced that replaced welding by hand. Machine tools were driven digitally. Tools were organized into cells and continuous flow manufacturing of product was introduced in an automated fashion. Automated material handling and warehouses were a key element of processes. They were built and later were dismantled or shifted to other service as firms realized there was no value in storing parts or materials. Automation made it easy to do but the cost of the inventory did not conform to making mass production into an automated lean manufacturing capability. The industrialization of manufacturing was not restricted to the discrete part of modern industry. Automation also happened in the chemical, pharmaceutical, and process industries. Materials were moved through pipelines. Control systems operated the plants to make consistent and quality product. In both the discrete parts and the process industries, material handling automation provided for a controlled operation and fixed routes for materials to travel. Human-free manufacturing systems and lights-out production floors were the vision of manufacturing operations. Lean manufacturing automated the entire process, with information, materials, and product flowing in a disciplined fashion along routes that had been previously established. Automation provided a manufacturing capability that was very productive, reliable, and reproducible. It made products with precision and discipline.

The concept of automation and agility can initially be challenging to understand. Since automation requires a rigidity or reproducibility within its operational activities, being agile and responding to unexpected change seem to be the opposite of automation. The agile athlete leaps and runs in what seem like random patterns. The agile manufacturing concept is thought to be of the same nature. Agility is sometimes envisioned as people rushing around to respond to an unexpected demand of a key customer. The atmosphere is crisis and the impossible is made possible. This is not nor should it be the case. Agile or custom manufacturing capability must plan for the unexpected change, and it must anticipate that it will occur. Capability, then, is built into the response in an agile but predetermined fashion. Providing the discipline for the agile or customization supply chain requires a degree of automation of the manufacturing process, the material movement, the information system, and the movement of cash. The understanding of how automation will apply to the agile or custom manufacturing supply chain is an important concept in the progress toward a more responsive supply chain. This chapter will discuss the automation of the machine or process and how automated capability can be developed that will make agile rapid response easier. It will also discuss methods of automation that allow for agile material handling as well as how automation deals with information and decision-making processes. The automation of the movement of cash in support of an agile supply chain will also be developed. Finally, the chapter will define the development of the discipline that an agile automated system must have to be effective.

The foundation of an agile automated system is knowledge about the product and process. Knowledge allows the building of a capability to rapidly respond to unexpected demands or other changes. Product and process knowledge is needed so that the product can be produced flawlessly. Product must be made right the first time if it is to be delivered with the speed that agility implies. Time cannot be spent testing and/or reworking. There should not be time to satisfy the customer with a second run of the product. This same knowledge is required to automate the machine and product to be made on it. This knowledge is what allows a capability to be developed that enables additional capacity or other machines to be brought on-line when demand exceeds what the primary process or machine can deliver. Having a predictable and reproducible capability to make the product becomes the center of the supply chain. The other parts are developed with the understanding of just how the machine, process, and product capability will perform.

An example of where knowledge of the machine, process, and product allowed for the development of supply chain capability is illustrated by a European manufacturing company. The company supplies elastomers to customers that use it in making intricately molded or extruded parts from liquid materials. The product is a two-part system that is mixed together by the customer to get the liquids to cure into a solid rubber material of the desired shape. The properties of the final shape are determined by the formulations of the two-part system. The formulation controls the rubber parts' extrusion rate, color, cure rate, and elasticity. These four properties are modeled relative to the formulation and the performance of the machines that separately compound the two-part elastomeric materials. The order is received with the final part properties defined. The model determines the formulation and it is sent to the machines that produce the products. The quantity desired and the package or container are also specified. The machines automatically produce and package into separate containers the required quantity and the performance is guaranteed. The model determines what will be fed to the machines and how they will be processed. Packaging is specified to meet the customer requirements. Both package and product are an example of customization. The customization is automated for each order. No finished-goods inventory exists, and the whole process is initiated at the time of receipt of the order. The reliable and predictable ability to satisfy the order is the foundation for a supply chain that operates in a streamlined and rapid fashion. Automation exists because of the knowledge about the products and the manufacturing system.

The reason this system could be developed is the knowledge about the relationship between the formulation, the operation of the machines, and the performance of the elastomer products. Some formulations specified in the orders have never been made before. The model was well enough developed, however, to guarantee that the products would perform. Other formulations are ordered frequently and thus are well tested. Each run is incorporated into the model, and the company's ability to specify formulations continues to improve. With all products customized for the customer, the entire supply chain can operate agilely and automatically. The other capabilities in the supply chain are consistent with this capability.

Knowledge is the foundation of automation of the machines, the material handling systems, and the information and decision processes. It is all based on turning that knowledge into a discipline that can be repeated flawlessly. Thus, the capability must exist to operate error-free in a fashion where the supply chain can be counted on to perform. Random upset must be managed or eliminated from an agile system as it is eliminated in a lean system. Root causes of statistically significant errors must be systematically eliminated and more knowledge gained on all parts of the supply chain. An agile operating supply chain is not a freewheeling creative drive to the basket by Michael Jordan. It is a predetermined response from a capability that has been developed and automated. An agile operating supply chain can handle all the variability in demand or other events. The development of the supply chain capability is based on the anticipation of unexpected events.

Agile Automation— Process or Machine

The automation of a process or machine can exist in an agile environment. The automation must be consistent with the capability of the entire supply chain. An example of automation would be a Michigan company that was in the business of making air drives for machines and laboratories. Its business involved purchase of castings and then machining to add vane rotors, silencers, and other components to the product. The company had been using a machine shop that was laid out by the type of machine or operation. Machined parts were processed through the factory with the different steps being performed at the different machine centers. Final assembly brought different parts together and built the air drive. Then the entire plant was restructured from a conventional machine shop to one based on continuous flow manufacturing using computer-driven agile automated machine tools. The process was capable of making product of all the sizes. The change replaced the standard machine shop layouts with two cellular layouts that performed all steps in the production, assembly, testing, and packaging of the air motors. The cellular layout was based on starting a product and taking it to completion in a continuous flow manufacturing process. The process started from castings of the bodies where various machine tool tasks were performed on them. These tasks were driven by an automation system that was preprogrammed to perform the right machining action at the right time to shape the parts that went into the air motors. As the parts were finished they were assembled, tested, and shipped to the customer. The directions and management of the cells came from a computer in the engineering area. Once the software and instructions were loaded, the process worked automatically. The only person on the floor was one who looked over the whole cell and also did the testing and packaging.

New or custom-made air drives took engineering development and programming of the cell. The standardization of the software instructions enabled this to be done efficiently in a minimum of time. The critical path to a new part was the definition and delivery of the new castings that made up the raw material for the cell. Capability existed in this cell concept to effectively handle unexpected shifts in demand for the standard product line and for the development of new air motor products in a customized and automated fashion. The cells were built to be agile and respond to unexpected change. In this case, the new manufacturing capability and its technology operated in only 10 percent of the space with many fewer people. The old part of the shop was idled. Inventories were reduced to less than half of previous levels, and customer service was greatly improved. Customized drives were introduced into the product line. A new prosperity existed in the firm. The new technology fit the supply chain and profitability was improved.

Providing automation to machines or the process as in the preceding example can be very technical and require lots of creation and discovery. Assessing just how and what to change and automate becomes a challenge for the people developing the improved continuous flow manufacturing line and integrating it into the supply chain. After envisioning how the supply chain is to operate, the developer must work on every step of the supply chain. Specific attention must be paid to the manufacturing steps. The manufacturing steps require the development of the activities that must be accomplished along with some standardization on instructions for how they will be done. These instructions must be agile to handle a wide variety of situations.

An example of this comes from Silicon Valley, where a custom manufacturer of circuit boards supports many of the companies engaged in producing electronic systems. Solectron, a company that assembles electronic components on circuit boards, can take a design of a board and turn it into product in 24 hours. This requires that the design fits a format that can easily be placed into the computer that drives the whole process. Once the standardized instructions are entered, the printed board can be produced automatically. The bare board has the pathways or circuits printed on it automatically, which are then fed to a machine called a "shooter." The shooter places the electronic components onto the right location with the right orientation. This step is done very rapidly. The shooters work so fast it is hard to see the movement of the placer's arms. Various-sized boards with different circuit printing on them can be fit into the process, and most of the components can be placed by the machines. Any special devices that go onto the board are done by hand. Besides specializing in highly efficient, long-run board production, Solectron also provides a smaller-scale operation for limited or initial production runs. It allows customers to learn from the production process so they can choose to make the larger runs themselves or to have Solectron do it for them. The processes are automated and the production rate is rapid. The number of boards automatically produced can range from less than a hundred to more than a thousand. Larger runs can be made on the more specialized process lines. The machines are automated, and the steps along the lines are managed by a computer. The assembly process is shifted from workstation to workstation with the final step being testing. The whole process is mostly automated with very high speed robots as placement machines. The process does have spots where manual operations are performed. This facility is a classic example of an agile manufacturing operation. It does not know what business it will have more than a day or two in advance but responds to that uncertainty rapidly.

Automation of the machines or process can be done to provide agility. Companies that do this the best will have a competitive advantage. In providing automation, continuous flow manufacturing must be used as the concept for movement of raw materials or components through the manufacturing plant and to the customer. With creative development of the capability, it can be agile, provide for customized product, and anticipate and respond to the unexpected.

Agile Automation— Material Handling

With the manufacturing process agile and automated, the supply chain also must be built to do business in this fashion. This requires the material handling within the supply chain to have the same level of responsiveness as the machines. If a supply chain is to be agile, it must have the capability to effectively move materials into the process, along the manufacturing steps, and away from the process and to the customer. This requires a degree of automation consistent with the materials being handled. The manufacturing process will dictate what this will look like, depending on the type and amount of material being handled. In the circuit board assembly of Solectron and Motorola the manufacturing processes are connected by conveyor systems with positioners. Various steps are performed, from shooting devices on the boards to soldering and testing. The entire movement along the process is automated and controlled. Loading the shooters is done manually with various devices on reels, which are programmed for selection and positioning. The boards themselves are inserted in the process and removed manually. Movement on the shop floor from one process to another is automated and integrated by a computer. Some materials, like the reels for feed to the shooters, are brought to the machine manually. The movement of the product through the system is in a continuous flow fashion. Only unused components are returned from the machines to a nearby storage rack on the shop floor, where they reside waiting for the next product that needs this kind of device or component. These systems satisfy the automation requirements for the material handling of these kinds of processes.

Other processes will bring raw material or components from a truck or warehouse in bins or on pallets. It is desired that the entire manufacturing process will be done at a location like a cell or process line. The product would move via pallets or container to trucks for shipments to customers. In other cases, material moves via pipeline to packaging equipment. After packaging, it is moved by conveyors to palletizers and shrink wrappers and then to the truck for shipment directly to the customer. The important feature of these systems is that they are not overautomated, and agility is maintained. They are responsive and intervention can occur, and custom or special situations can be handled with ease. Not every step is forced to fit but provision is made for predetermined manual intervention. The automation conducts or directs all the tasks. The capability must be built to fit the agile response desired but not overbuilt. A computer or developed procedures drive the process. Procedures do not need to be complex and fancy, but they do need to be thought out and perfected through use. Activities must be carried out in a reproducible fashion.

Agile Automation—Information and Decision-Making Processes

The heart of an agile process is the information and decision-making system. This is where the direction emerges to be able to handle any unexpected circumstance that arises. Various decisions can be automated with information coming from various sources, including the manufacturing equipment or process. The order that drives the system can be looked at carefully to determine if all the requirements are spelled out and appropriate. This then can be what drives the manufacturing process. For example, Solectron can create the design of a circuit board and determine the manufacturing process in about 12 hours if the information is provided in a formatted and complete fashion.

The order system can also automatically determine the ranking of the customer. This then determines how the customer will be treated in the manufacturing and delivery process. The top customers will get special attention, and their orders will be made and delivered as requested. The lower-ranked customers would be fit into the schedule when time allows. Developing the information system on how each customer will be treated requires establishing an understanding within the supply chain team that all customers are not the same. Once this is done, the information system can treat each one in the fashion that has been predetermined.

With order inputted and customer priority established, the system can do an analysis for available material and equipment to produce for the order. Software exists that is based on models that will work through sequences to eliminate bottlenecks. It will figure out how the order can be fulfilled. As the order is being made, some information systems can monitor the manufacturing variables or parameters and determine that the unit produced is correct. This can eliminate final product testing or inspection. Knowledge that the process was in statistical control and controlling the parameters properly ensures the quality of the product. This parametric acceptance, which was discussed earlier, began in the pharmaceutical industry and its use is expanding. The goal is to eliminate testing by managing the machines used to produce. The materials or components going into the manufacturing process must be the right ones and of the right quality and quantity. The variables or parameters in the manufacturing process must be controlled precisely, and the product must be packaged properly. All this is monitored by the system, and certification of product occurs. In an agile system, where the product requested is unique, this can occur if the process to determine how it will be made is predetermined. The system or computer can help with determining how a new product will be made but only if the capability to do this has been developed or designed into the system. Solectron has designed the capability to make almost anything that can be put on a circuit board. It developed the system to make this agility happen in an automatic fashion.

The materials that go into the agile process must be tracked using either visual inspection or scanners. Bar coding is a common method of bringing information to the computers and can be used to determine the position of product or components. The tracking system is a key element of the automated system. Besides knowing what goes into every product, it can measure the progress of the units via automated run sheets that are also computerized. The product can be tracked through the process and onto trucks for shipment to the customer. The location of the truck can be determined while moving to the customer and any adjustments made to the route or the pallets to be delivered. The products can be scanned by the customer and the receipt entered into the computer systems. The truck also can be released from the delivery with this scan and input into the trucking computer system. This automated receipt drives the internal information systems for the customer, the supplier, and the truckers. Payment could also be initiated upon receipt and the product can be made available for sale or further processing.

The automation of an agile process or supply chain can occur but needs the capability built in to be able to respond in an anticipated fashion to the unexpected change in demand or product. It has a foundation of knowledge about the product and process and the degree of variability that will be requested by the customers. This is then turned into a disciplined system that works effectively to deliver make-to-order products with a minimum of inventory in the supply chain. Knowing what will happen in response to the unexpected or the routine is the discipline that the agile supply chain needs. It can be done using a computer or other type of system. Automation requires a precision that comes from either following procedures precisely or having them directed by a machine like a computer.

Agile Automation—Cash Flow

In the automation of cash flow the desire is to minimize the investment in capability and to turn the cash cycle as quickly as possible. The drive is to keep the level of investment low and to turn products into cash rapidly. Minimizing the investment requires that the capability operate with as close to no inventory as possible. The value-added time divided by the total time that materials spend in the continuous flow manufacturing process must be as high as possible. Materials cannot sit idle and be in a non-value-added state. This is accomplished by having a supply chain capability that is agile and can effectively adjust to the variations in demands from the customers. It must be able to do this because all components of the supply chain work together driven by customer orders.

The utilization of physical assets is another important optimization. Automation of the process or material handling capability enables full use of the assets to satisfy the customer demand. The combination of investment in inventory and assets must be optimized. The benefit of less inventory can be used to increase the fixed asset capability and thus build more agility to respond to customers' needs. The candy example described earlier was just such a trade-off. The result would be a more satisfied customer. The entire supply chain would be more responsive due to the streamlining of the material flow and the automation. Cash flow would be improved.

Inside the operational supply chain is a cash cycle. It involves receipts from the customer and payment to suppliers and others. The typical company wants to receive payment from the customer before the suppliers are paid. This allows the operations to occur using the supplier's money. Sales and procurement terms form the basis for the cash cycle. Automation and agility can play a key role. Automation can allow the payment to be made electronically with the customer's receipt of the product. It can help in keeping the inventory at the customer down by just-in-time delivery or vendor-managed inventory. These are benefits to the customer but also to the supply chain as a whole.

The supplier also has a desire to be paid quickly and optimize the cash cycle. The same principles apply here as they do for the customer. Electronic payment on receipt with a just-in-time or vendor-managed inventory applies here also. With both the customer operating with a lower investment in inventory and with electronic commerce, the supply chain will operate with much less cash. The supply chain automation allows for optimization of the cash flow. The automation is applied to the transactional part of the operational system and the way business is conducted. Many of the benefits of automation come from improvements in cash flow and are thus the drivers of the shift to agility and mass customization.

Discipline in Automated Agility

The development of capability based on discipline is required to respond effectively to unexpected demands along the supply chain. Although discipline itself does not give agility, it ensures that a response is consistent when the same circumstance arises. The foundation of discipline comes from knowledge about the product and process, which can be built into a responsive capability. Without discipline, the supply system will operate in a freewheeling fashion; things will be forgotten, and delay and errors will be introduced. Corrections will take time and effort. During those delays the capability will sit idle or resources will be used to make things right. The cost will make a nondisciplined agile operation less competitive. Discipline in a system will allow for making it right the first time.

There is a temptation to make something agile by investing in resources that wait for the order or unexpected demand to appear. In a competitive world this does not work. For example, a specialty chemical supplier developed a make-to-order capability with delivery in no more than three days. It built a capability to respond in a disciplined fashion for the last couple of process steps in the manufacturing part of the supply chain. These steps added the catalyst and pigment and removed particles from specialty chemicals. The chemicals were also packaged to the customer specification. The process ran successfully but required a very large inventory of raw material or intermediates to support the operation. That large inventory was produced in long runs from batch equipment. The intermediates were tailored to the final specialty chemical process. A number of these would be used in each specialty chemical as it was finished. The chemical intermediates were made to internal standards and the product line had 70 chemical intermediate formulations. All were developed over time to provide specific properties when the company was rewarding people aggressively for new products. Rationalization of reducing the intermediates to less than 20 had not been successful. This attempt at agility did not provide the market a product that was made competitively. Profitability was a problem. The manufacturing system lacked discipline. The intermediate production was not integrated with the finishing process so that materials were replenished in a just-in-time fashion. Instead, long runs were tied to a planned need. They were not linked to a demand or pull need. The lack of discipline in product line development was very apparent with a reward system that encouraged innovation and new products. New product sales became an undisciplined driver that affected the effectiveness of the supply chain.

This specialty chemical manufacturer has a specific challenge. It must build a disciplined approach to the product line and narrow the number of choices at the intermediate level. Another choice would be to discipline the system by driving the customer's order further back in the supply process and integrating all the process steps with pull demand from the customer. The customer order would initiate the production of the intermediate chemical as the feed for the final steps in finishing the specialty chemical to the customer's need in a customized fashion. The changes will result in a more agile response and also allow for new product introduction into the whole process. The supply chain lead time will be reduced and the investment in inventory will make cash and expense available for other uses.

Clearly, automation and discipline are essential to developing an effective agile operation. The response to unexpected change must be preplanned and put into a system that has an expected and predictable response. The machine or process must have well-thought-out agile automation and be integrated into a material handling system that is balanced with the needs of the agile machines and the supply chain. The information and decision-making process must also be automated so that results are reproducible. The whole automated supply chain must be operated in a disciplined fashion to achieve the desired level of competitiveness.

This chapter discussed the issue of automation in modern industry. The next chapter deals with new processes, products, and markets. It will attempt to provide an understanding of how a company must react to the different degrees of complexity that are inside the opportunities that drive new growth. The most complex situation exists when one is trying to be the most bold and creative.

PART 3
Changing Capability in a Modern Enterprise

CHAPTER 8
New Capability—Products, Markets, and Processes

New product, market, or process capability is the lifeblood of industry. It is where research and development provides the new technology that results in improved commercial performance of the firm. It is where the success of the new technology commercialization effort is ultimately measured. It is where regional and global expansion of markets occurs to broaden a firm's presence. The ability to sustain or generate revenue growth or improved profits is the test of new capability success. With the vital nature of this activity it is important to understand how the commercialization process works and to determine those things that make it a success. The commercialization process must be agile and able to respond as the circumstances of new products, new markets, and new processes are explored. New capability must also develop technology that can provide for agility as it is introduced and faces the test in the marketplace.

The Three Dimensions of Commercialization

The three dimensions of new commercialization deal with product, market, and process. These dimensions need to be dealt with individually or in combination, depending on the circumstance of the commercialization. Just what new capability is needed will determine the complexity of the effort. It will also determine the level of risk that is involved. A commercialization that involves building new technology and capability for product, process, and market is more risky than simply taking existing products into a new market. Both have risk, but one that involves more than one of the dimensions will require more capability to be built and thus have more opportunity for error or failure. Understanding and managing the complexity is a major challenge for the commercialization process.

The dimensions of commercialization can be shown in terms of a three-dimensional cube with each side being product, market, and process (see Figure 8.1). Each dimension has the current, or existing, capability at the origin point and a new capability along the axis. This model defines the commercial world that new technology or capability will encounter as it is developed into expanded business. The Dimensions of Commercialization cube defines the challenge and the complexity that the process will encounter and shows the interaction between new product, market, and process. Existing business is in the upper-left corner. This business is in a state of continuous improvement with delivery to customers as the focus of the operational enterprise. All other parts of the figure represent new commercial activity.

New commercial activity can be anywhere on the Dimensions of Commercialization cube. The amount of activity required to commercialize depends on where the activity rests on the diagram. The most complex technology development is in the lower-right corner. Other parts of the cube have varying degrees of complexity. Introducing a new product into an existing market from an existing manufacturing capability can be straightforward. Launching a new product into a new market with a new process (the corner opposite existing business),

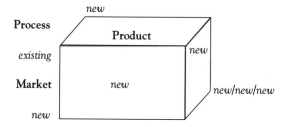

Figure 8.1 Dimensions of Commercialization

however, can be very complex. If all three dimensions require new technology, then the change is really complex and special attention and agility are needed to make it a success.

One could introduce a *new or modified product* into an existing market using the existing process to produce it. In this situation, only one of the dimensions is changing, and the change is one of the least complex activities of the commercialization. It is also one that needs to be managed very carefully. These new products tend to take over existing product sales; thus, the need to refresh the product line should be incorporated into the decision process. In many cases, easy internal reward for product developers comes from this type of activity, and the tendency may be to refresh the product line and expend the resources to do more than is necessary. An example of this would be providing a new and improved soap or detergent. The formulation of the product may change and the package may have new graphics, but the process to make and package the new soap is the same as existing business; the market, geographic region, and channels of delivery are also what are being used today. That does not mean that this process always works as effectively as it should.

In 1992 *Fortune* magazine ran an article titled "The Dumbest Marketing Ploy." This example, which was discussed earlier, depicted a promotional activity at Procter and Gamble that was being changed. The chief executive officer Edwin Artzt referred to what they had been doing as a monstrosity. The change showed the impact on the supply chain of a strategy that dealt with deals and promotion. The deals and promotion were the heart of the way the industry operated. This approach was aimed at gaining market penetration for P&G's product offering. It dealt with the concept of new and improved soap and detergent products as a part of that promotional mentality. Many changes were in the package or the claims but involved the same products. The results did not make P&G the company that the new leadership felt it could be.

By dealing with a promotional and deal approach to the sale of the product line, P&G required the supply chain to be prepared for the surges that resulted. The manufacturer had to stockpile the ingredients and prepare for huge runs of existing and improved products. Workers required overtime to produce, and temporary workers needed to be hired to meet the surge in production. Freight companies charged premiums for the manufacturer's periodic blowout shipments. Distributors overstocked as they binged on short-term discounts. Product sat in the warehouses. Goods were overhandled and damage occurred. The consumer bought goods that were old and freshness was not assured. The process was not streamlined or efficient. It cost a lot of money for the promotion and deal approach. The capability was built for the peak demand of the promotion and deal. Product sat idle for much of the time. The grocery shelves were flooded with new product for which there was no room. Discounts were needed to move product. This process is known as "trade loading" and is practiced by many industries.

The new leadership of P&G leveled out the flow of goods through the supply chain. This resulted in the freeing of cash by not having to stockpile raw materials. The workers were not exposed to surges of activity. Distribution and warehousing processed the average flow of existing and improved products. Excessive handling was eliminated and reduced damage resulted. The product arrived at the store shelf in fresh and good condition. The result was a cost to the consumer that was 6 percent less than the promotional and deal approach. The new concept of doing business allowed for the introduction of new and improved product within the context of a streamlined supply chain. The approach involved changing both the products that were offered and the processes or supply chain that got them to the market. It involved two of the dimensions of commercialization.

This type of industry, because of the large investment for peak loading of the system, has a fair degree of agility. By providing for a smooth flow of goods to the consumer, the needed capability can be understood and the remainder can be idled or used in another fashion.

It can be used to develop products with the channels and manufacturing and packaging processes well understood. Focus can be on understanding the changing demand and new product needs in the marketplace. Competitive activity can be responded to in a rapid fashion because of the excess capacity in the streamlined supply chain. The strategy can be one of both proactive and responsive product introduction. The result at P&G is not only a more streamlined and efficient company but also one that has more capability available for new commercialization.

The P&G example highlights the need to be operationally effective but to also have the capability to agilely respond to demand or needs that arise in the marketplace. Companies that service the consumer must understand how their offerings fit in the competitive situation. They need to draw on the wealth of knowledge they have about customers' needs and their potential product offerings to ensure they are providing what customers want. The wealth of product technology can be tapped to meet changing consumer needs. This exists whether the industry be soap and detergent, cigarettes or chocolate, or snack foods. The leaders in the industry must know their product technology for both their existing and future offerings. That product technology includes both the process to produce and the technology to package and deliver something that will appeal to the consumer. Continued success in the marketplace requires that the typical consumer company have an agile capability relative to bringing new offerings to the consumer.

A *new or modified market* dimension presents a unique challenge to a company. Existing markets are normally well known to the firm. They are their lifelines to success. Bringing an existing product to a new market or application using the existing manufacturing capability is not an easy task. It requires establishing relationships that have not existed. These relationships have to be developed into ones that can be relied on. Understanding the new market and the customer needs presents a real challenge. In many cases companies assume a product will work because it is the best technology and fits the needs of the customer. To their surprise the market does not react as expected, and the introduction into the new market fails. That failure is usually a result of not understanding the details of the market and making too many assumptions that should have been tested. The existing product aimed at the new market must work in the new application and meet the needs of the customer. The supplying company must be able to deliver it under the conditions that are required through channels that may be new. If the application is successful, then a surge in demand or a need to improve or adapt the product may result. Both these needs must be handled by the capability that exists in the supplying firm. This firm's capability must be able to respond to demand, even when demand is anticipated. Often firms are surprised when what was hoped for really happens and the capability is overwhelmed. Failure to deliver at this stage can result in failure in the new market.

Introducing an existing product into a new market requires that agility be developed to ensure the supply is available. For example, plans must be prepared that include using other equipment or excess capacity if the demand is realized. This equipment and capability may exist outside the firm and should be reserved at the time of introduction. The product for the new or modified market also must be qualified on the new equipment in anticipation of demand that will exceed the capability. Success requires the development of an agile response.

The Dimensions of Commercialization cube also has a *new or modified process* to manufacture the existing product for the existing market. Change capability in this realm can be simple or very complex. In some cases the new process replaces an old process, but the improvement does not involve a significant change in the product. Other situations will involve a complete change in the way something is produced; thus, customers must be notified and qualification of the new process will result. The drivers for process change are many. Perhaps replacement parts are no longer available for the machine that is being used. Or maybe a new machine has been introduced that significantly improves the cost to produce. Or capacity may have been exceeded by the growing demand and new capability needs to be brought on-line.

There are also other drivers. For example in the chemical industry, the process may require new chemistry because the existing capability has too large an environmental impact. This then results in a significant capability change. Sometimes it may even mean new raw materials. Another driver may be that the market has shifted out of the country or to a new area and the new process follows in order to ensure the business. Global responsiveness is required for success in many industries.

So far the new or modified capability with regard to one variable or dimension at a time has been discussed—either a new or modified product, market, or process. Things get much more complex when we talk about more than one of these capabilities needing to change at one time. The most complex scenario is when all three dimensions change. This is depicted in the Dimensions of Commercialization cube as the lower-right corner. Bringing a *new product* into a *new market or application* with a *new process* is a very high risk endeavor and requires special management attention and skills. Coordination of this type of change requires people with various specialties. They may be technologists who specialize in the product or process. They may be people who focus on the future needs of the market. The combination of these various skills makes this type of commercialization much more challenging than a simple new product introduction.

The people skilled in understanding the needs of the customer must make sure that the product satisfies that need. The people with skills in getting the product into the market will need to be developing the relationship with the potential customer, often with no previous contact and without a relationship that can be built upon. It also must be done without detailed knowledge of the potential customer's application. The people with skills in developing a new manufacturing capability will normally doubt whether the venture will be a success. Typically, they will undercommit effort, and the processes will not be ready when the product and market need them. This is a natural response, since the investment in the process is the first component that puts significant capital at risk. The required skills and resources that are needed to develop the process technology are held back until there is proof that the business will exist. They have seen too many of these types of commercializations fail and are responding accordingly. Getting the commitment of all the necessary people and resources requires a special treatment for these types of ventures.

Examples of these types of ventures are many. Some meet with significant success and reward; most do not succeed. Many of those that do succeed do not meet expectations. Time to commercial success is usually much greater than anticipated. Staying power is challenged.

New materials for the defense industry are classic. The industry pushes the envelope of performance and thus needs materials that do things that have never been done before. Success on a prototype does not mean that material will be ordered shortly. It just means that the defense industry has seen a new material that looks good and it is searching for the application that needs it.

A supplier to the electronic industry recently developed a completely new material for device manufacturing. It performed exceptionally well and gave a large advantage to the companies making chips and boards. It was a high-performance product with electronic grade purity and sold in pint bottles for thousands of dollars. The customer that developed the product in concert with the materials supply firm accelerated the demand curve. The material supplier responded in a very agile fashion by developing plans to put two facilities up in less than a year, one in the United States and the other in Asia. The capital cost was around $70 million. Because of the acceleration of demand or the lack of being prepared, the materials supply firm had very poor process technology. It had a hundred pounds of waste per pound of product with a very low throughput in the pilot plant. Resources were immediately increased 10-fold to make the process viable. Work was proceeding and was successful. The construction of the plants to be completed within a year was still on schedule. The buildup of personnel was yielding results. Then, the customer rethought the program that these materials were a part of and reduced the demand to a lower level. The firm supplying the material had responded agilely to the upside. But in the definition of agile, it had not anticipated and prepared for the unexpected. It did gear up and was ready as the demand dictated. With the drop in demand, the question was "What to do now that the demand has been lowered and a less rapid pace is needed?" The firm stopped the capital projects and reassessed the capability that could be gotten from the pilot plant. They shifted people back to their old jobs or on to new ones. They kept the process people working. They had learned enough to now supply the lower demand from the pilot plant.

The internal reaction to this situation was for people to complain about all the stress that had been caused. For example, internal organizations or projects had lost people or capital as the agile shifts were made. These organizations were not happy that they were being yanked around and disrupted for something that did not happen. A respected senior engineer in a high-level position broadly circulated a letter that congratulated people for their agility and logically convinced the unhappy people that this was the way business should be done. The responsive techniques needed to be a part of the firm's tool chest. Today, the product is a success, even at a lower demand, and much material is being sold. The product is growing at a rapid rate but now more within the capability of the organization. However, the economic ramifications are great and the stretch of the pilot plant has resulted in significant delays to the commercial plant. The pilot plant has enough capability to be the second source, but the Asian investment has been delayed for an extended period. The point in this case is that agility is needed both for the upside and also when a downside occurs.

Other companies have managed these types of ventures with success. Two examples of success have something in common. Motorola's early work with cellular phones was treated with skepticism inside Motorola. Optical fibers were being given the same treatment inside Corning. Both companies attribute the success of these new products and new markets, with new processes, to establishing the project directly under the top person in the firm. These innovative concepts required protection from the skeptics and then some very bold actions. The existing organizations could not find the resources to support the bold actions and did not have the will to make something like these creative and totally new ventures work.

This type of commercialization requires creativity and a degree of boldness or risk taking beyond what is required during normal business. The product must be developed for an application that might not be the place where it is eventually used. Manufacturing capability must exist for quick prototype production of the material or the device. Additionally, the plan for rapid deployment of manufacturing capability must be developed with agility. Unanticipated change is the rule, not the exception. The product may change many times until it finds the application. The process to produce the changing product must also adapt to this change. The demand may vary depending on which application a particular customer hits. The success of the finished device or product of the customer is also a significant variable. Did the customer's customer really mean what was communicated about the need? Developing the ability to respond in an agile fashion in this domain is a requirement. The firm that can accomplish that will increase its probability of success. Other firms will avoid the new product, market, and process domain completely, realizing that the risk is beyond their capability.

A new product, market, and process activity that also involved new science and technology occurred recently at a firm in the materials business. This company chose to explore the development of some high-performance ceramic materials. These materials were made using a completely new process technology that started with new chemicals and went forward to fired ceramic parts. New science was necessary at almost every step. This effort took almost 15 years and a lot of development money. The benefits from the business that exists today are not large enough to pay for the cost of all the effort, however, even with complete technical success. The question that needs to be asked is, "Would being more agile have helped this process?" That is, when the critical path to success is technology discovery, will it pay to be more agile? One could put forth the premise that being agile in forming partnerships would have brought more talent to the technology development effort. It is very important to be looking at different technology routes to potential new opportunities. This agility may not result in technology choices, but in many cases it does. Then the commercialization team must determine which courses to pursue. Discovery is not always a predictable event, and agility is desired but not always attainable. Undoing the puzzles that Mother Nature provides sometimes is not as easy as we would like it to be.

The other parts of the Dimensions of Commercialization cube have various combinations of change involving two of the dimensions or variables. These will require a much more intensive effort than when only one variable is changing as commercialization occurs. Making sure that capability is developed appropriately is essential to success. Introducing a new product into a new application using the existing manufacturing capability is a common event in industry. The focus is on customers and their needs. The product development people strive to provide the right product capability that satisfies the requirements of the application. The process person knows that the product can be produced. It will require effort to make it a part of the production capability using existing equipment.

A new product produced in a new process makes for a slightly more complex situation as the investment in process technology and ultimately physical assets requires a more significant commitment of a firm's resources. Whether the product will result in the expected business is a major question and becomes a part of the decision to proceed. The key to making this an easier decision involves the use of agility. If the process can be developed so it can be used in other applications, various avenues to success are possible if increased business is not generated by the new process. Providing a production home for the new product and its potential family of product also is a major concern. The production home may require new process technology and equipment. Again, the risk of the commercialization is lessened if the new equipment could be used for existing product growth if the new product and its potential family did not succeed or had a lower growth than expected. This approach does require the development of agile capability that can serve more than one master.

A new market or application frequently requires a new process. This occurs when a market or application exceeds the capability that is in place to serve a much smaller market. For example, the product may be moving from a specialty application to a high-volume application. Penetrating the new application requires new production economics, and thus a new capability is needed. The risk associated with the process is somewhat reduced because knowledge has been gained on producing the product, but the risk associated with the new application regarding whether it will develop as intended is the same as for a new process. A plan that commits capital assets to a process that has no other use may not be the right one. It should look at the new application as being a success and build appropriately but should make sure that the process unit is not so specialized that it has no other uses. Agility should be the watchword. With the emotion associated with the high-volume application and the business being "a sure thing," it is easy to get involved and committed. One must be cautious so that involvement does not result in a very focused plan without a customer. A little clearheaded planning with agility in mind can avoid the situation. For example, if a new market or application will result in a shift in manufacturing capability to high volume, it must be determined whether the low-volume process technology will work. The production at expected high volume will make the flaws in process technology much more visible. Making the product right the first time may not have been a requirement at lower throughput. The amount of waste per unit of product may not have been important with the existing levels of production. Change must occur and improvements selectively made to the manufacturing capability. Another driver of change to the process and supply chain capability can come from the market. High volume may require a different way of doing business to reach the success level desired. Customers may require a degree of personal attention, and mass customization may then be the direction. This requires a different process and supply chain capability than was developed for the lower-volume product. It may be that the primary physical nature of the product is the same but it has changed so that each customer feels it is being made for them. Mass customization needs to be developed.

Mass Customization— A Part of Commercialization

Today's customers have become more demanding than ever before. They are looking for variety and uniqueness in what they buy and the services that they receive. Moreover, they want this variety and uniqueness as well as high quality and low cost. Customers also change what they expect with great frequency, depending on what the marketplace offers. The supplier is challenged to satisfy these changing and demanding expectations.

For firms to meet the expectations of their customers, they have had to change the way they do business. As discussed earlier, new products from new processes must be introduced into new markets to provide the variety that customers expect. Mass customization is also a technique to provide the uniqueness and the variety that the customer is demanding. The diverse needs of the customer must be effectively satisfied if a firm is going to enjoy market growth. Companies must meet the customer need for variety but not lose sight of the need for excellent service, high quality, and appropriate cost.

Hewlett-Packard (HP) developed a concept of commercialization for a new printer market that has become the standard of the industry. This venture involved all three elements of the commercialization process: A new product went into a new market with a new process to produce it. This added risk to the commercialization, but HP knew that all parts of the customer's expectations needed to be satisfied. Thus, the company set out on a path of mass customization. HP knew that the customer order needed to be filled ever more quickly and the need for low cost and high quality required a different approach. These customer-based demands had been confronted in a number of HP's businesses, including medical products, computers, and printers. HP has proven that it can deliver a customized product quickly and at a low cost. It is important to stress low cost. Many companies have attempted mass customization without the total success. That is, they have offered the variety quickly with high quality but with an effectiveness that made costs rise to unacceptable levels. HP has met all the expectations of the customer with excellence.

Hewlett-Packard's key to mass customization is effectively postponing the differentiation of the product until the last possible point in the supply chain. The benefits of this come from the ability to streamline the supply chain and maintain as low an inventory as possible. This is possible because the inventory is not committed to any product until just before delivery to the customer. These lower inventories along with effective design and manufacturing techniques make the supply chain able to deliver a high-quality product quickly and with the variety the customer expects.

Three design principles form the building blocks for this mass customization program. The product must be designed so that it can be easily assembled. This is done by making modules that have use in a number of products. They become functional components of many finished products available to customers, which provides the variety in choice that they are looking for. This must be done while maintaining the right costs and manufacturability of the product.

The manufacturing process must build the high-quality and low-cost modules that will assemble into the products with ease. The manufacturing process itself must also be able to be broken into steps or modules. The manufacturing modules must be able to be moved or rearranged and must support the different supply chain designs. They must be supportive of the concept of postponement of product differentiation yet be effective in producing and assembling the various modules that will be configured into the final product. Quality must be at very high levels. Plugability must exist in the product modules, and the tolerances for the fitting of the modular components must be very high. Meeting all these requirements is not a simple task.

The supply chain network is a very critical element of a mass customization approach to markets. The positioning of inventory as well as the manufacturing modules becomes a key to success in the marketplace. Locations for manufacturing and distribution facilities must be assessed to determine how quickly the market can be served with what variety of product. What is done at each step in the supply chain must be well thought out with the concept of postponement and low cost in mind. Part of the supply chain must supply the basic modules that will be assembled in the wide variety of products that the customer will choose from. These facilities must be very cost-effective with superb quality. The supply chain and distribution system must also be able to take the product modules and assemble them. This must be done agilely and with sufficient speed to satisfy the customer. The customer expects make-to-order capability of customized products that perform, are priced right, and are of high quality.

The module product design is at the heart of the HP system. Providing laser jet printers that work in all parts of the world becomes a real design challenge. Standardization requires compromise in the design process. It also requires real knowledge about the impact of costs as standardization occurs. Deciding to make the power packs globally standard could be looked at as a large increase in cost if one only looked at the cost of the component. When looking at the impact of having multiple modules for the various electrical currents that exist around the world, one realizes that the inventory items will increase significantly and the cost in the supply and distribution chain will be higher.

This provides the benefit that makes the selection of a universal power pack a cost-reduction decision.

Hewlett-Packard brought all the change agents of the printer business supply chain together to develop the agility required to satisfy the market in the most effective way. It was a collaborative effort of product design, process development, manufacturing, distribution, marketing, and engineering people. The integrated effort resulted in a make-to-order business with the final assembly occurring in the distribution centers close to the customer. The supply chain was very competitive. Suppliers saved on inventory with the standardized plugable modules. They reduced the transportation and duty cost to get the printers to the market globally. They reduced the time necessary to satisfy the customer with a custom-produced unit. They provided the market with variety and custom tailoring while significantly increasing the return on assets of the business. The build-to-order approach at the distribution centers put them ahead of all the competitors and established this approach as a best practice that will be emulated by other business segments inside and outside of HP.

Not all moves to mass customization require a new/new/new approach. Anderson Window, discussed earlier, modified its ability to take an order and integrated that with the production process to deliver custom-manufactured windows with less or the same cost as standard ones. Anderson's approach was not to take a bold step but one that implied entering or creating a modified market for unique windows. This required the development of a way to take and confirm orders that the plant could produce. The production process was also modified so the unique windows could be fit into the existing manufacturing capability without much trouble. The product changed its shape but was not different structurally than the standard product. Most of the technology that the custom manufacturing capability used was modified, not new. The new element was the ability to take a unique order in a disciplined fashion and to translate that into a manufacturing language that resulted in the unique product.

The shift to custom manufacturing can use a complete new technology and involve a change to a new process and a new product for a new market. This was the case with Hewlett-Packard. It also can be less bold and involve a modification of the technology so that custom products are achieved with less risk. This was the approach that Anderson Windows used to enter the custom window market. Just how much change or risk is required can be managed and the custom product can result.

Summary of Three Dimensions of Commercialization

Bringing new technology into the marketplace that involves development in the three dimensions of commercialization—product, market, and process—is a significant challenge in integration and collaboration. It requires that all three elements are successfully developed for the venture to be a success. The timing must coincide, because any delay in an element delays the program. The commercialization team must work together for the good of the product line. It cannot be fragmented; success of the venture must be the goal. There cannot be other agendas.

To make the commercialization process work, companies need to understand how bold or creative the individual dimension activities are. The team needs to understand where on the Dimensions of Commercialization cube the venture resides. If it is a new product, new process, and new market, then very special care must be taken to bring all three together at the same rate of development so that product introduction from the new capability can occur in a coordinated fashion. Each element must be strong enough to support the venture. The honest use of this tool can help to make sure the team understands the challenges of the commercialization opportunity.

This chapter has discussed the various aspect of adding new capability to an enterprise. It has presented a tool (the Dimensions of Commercialization cube) that can be used to determine if the course of actions is appropriate. When significant assets of the firm are being committed, agility must be a part of the strategy. If the investment is in a new market, there needs to be more than one reason to develop the relationships required. If the venture involves a new product technology, there needs to be more than a single application or product under consideration. When process technology is developed, it must be looked at as developing a capability that can be applied in a multitude of situations. Likewise, when process assets are installed, multiuse should be considered. Mass customization is a concept that involves all three of the dimensions. It is not simple, and a company's planning and thinking must address the complexity. The key to success is considering agility in the process of developing new capability.

In the next few chapters the topic of new manufacturing capability, new product introduction, and new market or channels will be discussed. Concepts will be developed that will improve the odds for success. Pathways for implementation of both mass customization and agility will begin to emerge.

CHAPTER 9
Processes

One of the dimensions of the Dimensions of Commercialization cube presented in Chapter 8 is the process. It is a critical part of the cube either as the only dimension changing or in combinations with market and product. This is shown in Figure 9.1 where the process dimension is emphasized.

The technology to manufacture either an existing or new product is critical in the successful commercialization. This capability, along with the ability of the product to perform, determines the success of the commercialization. The decision on the right process technology is not a simple one. It involves consideration of the customer, the supply chain, and the state of the product and process technology. Even when the product is an existing one and a new process is required, these factors come into play. The new process usually is needed because the original process is not meeting expectation; that is, it is deficient in some way. It may not have enough capacity to meet the present demand. It may not meet the quality requirements demanded by the customer. It may have an undesirable environmental impact. It may be in the wrong part of the world for where the market has moved. These and other reasons define the need for a new process to produce an existing product that sells in an existing market.

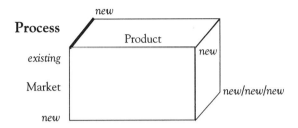

Figure 9.1 Dimensions of Commercialization

Process Dimension of Commercialization

More complex situations exist when either or both the product and market are new. If all three dimensions are new, then the venture is very bold and special attention must be given to this type of commercialization. It must be looked at as an opportunity where an individual enterprise may not have all the required capabilities. The manufacturing firm may have the process and product technology and capability but lacks the ability to get into the market. If the firm thinks that market understanding can be easily attained, it is probably underestimating the difficulty.

A company in the southeast chose to go into the fabric structure business. This was everything from tennis courts and golf driving ranges to greenhouses. It also included shopping centers and architectural buildings. The firm had a super product and an adequate capability to produce. The production capability came from fabric weaving and a new coating machine technology. This was followed by a cut and seal or glue facility that was ready for the big jobs that were expected to come. Each structure was a custom product needing architectural design. Even the common greenhouse needed to look good and needed a modular design to fit any situation. The fabric transmitted 90 percent of the light that struck it and thus had potential application for places where plant growth was desirable. The greenhouse installation was tested at an eastern university and was successful. The fabric also resisted ozone attack and stood up very well in sunlight. But instead of focusing on these properties for initial applications, the firm decided it would go into the more glamorous part of the business. It chose to compete with Teflon-coated fiberglass in the shopping center, stadium, and large building market. Each application was architect designed and was a custom structure. The architects continually expanded the challenge of the designs and expected the supplier of the fabric system to be able to achieve the new look. Customization was the challenge, but the capability did not exist in the company to effectively deliver the product, an installed and weatherproof structure. The firm was well out of its capability. Boldness led to failure. The firm did not understand this market and was drawn into the liability of the construction industry in a way it did not control. Faulty installation caused fabric failure. Liability of the installation caused the company to fail. To have succeeded, the company should have partnered with one or more companies that were already in the business, taking a slower and more controlled approach to the market. The firm needed to gain experience and knowledge about the product, the new process, and, most important, the market. Expertise needed to be developed. This was a case of arrogance, derived from other successes, standing in the way of clearheaded judgment—an example of not estimating properly what needs to be learned or known before one enters a business.

This example was a new process, a new product family, and a new market. The firm emphasized the process technology and put most of its effort into making sure it could deliver the structures that would perform. This decision, in the end, became the cause of the failure. The marketplace was not understood enough to make the right kind of decisions. As mentioned, when experience and learning are needed in all three areas, commercialization takes on an added level of risk. Failure that could sink the venture can occur in any of the three. The risk is not just three times that of a simple product or process commercialization. Because the commercialization is complex, and the people resources are typically limited, not everything can be worked on to gain understanding that would mitigate the risk and point the right direction. Special attention needs to be paid to these situations and agility must exist as the direction changes. The manufacturing process must not restrict the options that the venture has to commercialize.

The commercialization of a new process to make an existing product tends to be less difficult but is still not simple. Replacing a manufacturing technology that has been producing for some time means that the new technology must be better than the present process. Usually, some additional reason must exist for the replacement to occur. For example, typically, the current process has moved down the learning curve and significant optimization has occurred. In many cases the company's money is sunk into the existing facility, which is probably fully depreciated. Replacing an existing operating facility is a significant task. It is infrequent that economics would favor replacing the existing capability with new technology without some other factors entering into the decision. Replacing an existing process with a new one and having better cost and capital expectations put a significant burden on the new technology.

The need to replace a manufacturing capability occurred at the West Mifflin metal fabrication plant of General Motors. The achievements were presented at the 1994 Agility Forum National Conference in the winter of 1994. This was done with Alphonso Hall, plant manager, bringing his key people from the team to explain how they accomplished the redoing of the stamping process. Hall was credited by Brian Moskal of *Industry Week* as "Mr. Agility" in an article that appeared on May 15, 1995.

The plant was a part of the Service Parts Organization of GM. It produced the sheet metal parts needed as replacements for out of production body parts. Its business was driven by the collisions of GM automobiles. The unit specialized in small runs of parts like hoods, quarter panels, doors, fenders, and other stamped or welded fabrications. The product line had 700 parts and components and was growing. Each product required a fixture and a die.

The manufacturing unit was not competitive, however, and large inventories of parts were needed because of the 24-hour change time for the metal presses. This made the longest time not the run time of the part but the changeover time. It was analogous to what Toyoda found when he explored making cars in the late 1950s. He could not afford a press for every part like the assembly lines of the new car businesses of Ford, GM, or Chrysler. The GM plant had historically developed its capability as a stamping plant that serviced new car production. It did not change its operation when it shifted to short-run replacement parts and thus became less than competitive. It was being looked at as a shutdown candidate.

The West Mifflin team realized their situation and set out to change things. They needed a new process capability that could be changed rapidly. On analysis they found that the dies for stamping the sheet metal were stored outside and much of the change time was spent rebuilding the dies. A new process capability to store the dies in a managed and protected building was needed. A shut-down part of the steel industry served the purpose, and the dies no longer were the problem.

The improvement team next focused on how the dies and fixtures were changed. The plant invented a unique modular fixture scheme that utilized a common gridwork base plate with part-specific holding details that snapped into retainers. New gigs and tools were machined in-house and stored for each part. The system was built so that a forklift could push a die into place and make the change in a relatively automated fashion. The climax of the effort occurred in front of all the plant people when a die was changed in 10 minutes from a previous change time of 24 hours. The plant had truly achieved the ability to make one unit if that was required. As it turns out, dies are changed frequently during the day and the parts are made-to-order as the need occurs. This new process capability made the West Mifflin plant of GM competitive and set the standard for this type of operation. The process change involved a lot of new technology and cooperation among the plant employees. It was a success and the cheer that went up from the employees when the die was changed in 10 minutes demonstrated the importance of the new technology. The plant is no longer a candidate for shutdown. This example demonstrates that process change should be made for a reason. It is not always because a company is out of capacity or needs a whole new process technology. The need for an improved capability can drive the decision.

A new process is usually commercialized when new capacity is required, and then both processes are run simultaneously. It can also occur when opportunities present themselves in other parts of the world and local sourcing makes sense. The new process will take care of deficiencies that the old process has overcome by learning from running and it will have the potential of being much better as experience is gained.

After the decision is made to design and build the new process, the plan can be put together to bring the unit into commercial operation. The commercialization process probably requires that the initial product from a new process be qualified with the large customers. Because customers will be hesitant in making the change, they need to be assured that supply from the existing process is available if the product initially does not meet expectations.

This is not taken lightly and has resulted in significant delays in the operation of new plants. An additive for soap manufacture in Europe was being shifted from one site to another with an increase in capacity. It required the qualification of the new material at all the customers. The largest customer had the product fail on 1 of 12 front-loading test washing machines. Each machine was a different type, and the new process had to produce product that would work on all the machines or it would not be qualified. The type of machine was no longer produced, so the people from the new plant had to go into households and find the same type of machines. They replaced the old machines with new models. The tests then could occur to qualify the product. This was done and qualification efforts at both the new plant and the soap manufacturer continued. Supply continued during this time from the original facility. This delayed the introduction of some other products aimed at other soap formulations. These formulations were potential new products for competitors of the largest customer. In the end it was not clear whether the product was not qualified or price and market share were the customer's driving issue. Today the customer is very happy and the plant is running well. The other customers are getting custom-made products. Some are sourced from the new plant. This successful new process commercialization points out that frustration can occur and the time cycle in getting up and running may be longer rather than shorter than expected.

Procedure for Change — New Process Technology

New process technology usually goes through phases. These are similar in most industries. The first phase is to determine what it takes to make the process more competitive. At this stage a desired result is determined, and the existing process is viewed to see where it is deficient. This gap analysis starts the search for improved technology concepts that can be incorporated into the machine or process. This is followed by narrowing the initial list of the technologies to try. The next step is to build a miniplant, in the case of the process industry, or a prototype process, in the discrete parts industry, to learn more about the critical elements of the new process. Each investigation will focus on the areas of the new technology. It will rely on the proven technology to be integratable with the new at a later time. After the proposed new technology is perfected, with possibly recycling back to look at other technology, a pilot operation can be built on a larger scale. These units usually have the capability to produce the product in limited quantity, which is usually enough to show the customers. After the pilot plant has gained operational experience, a new process can be built at the commercial level. The pilot plant usually continues to run during this phase and is only shut down after the new process has been qualified with the customer, and the learning has shifted to the new plant.

The sequence just described takes a long time from start to completion. Significant effort goes into shortening the time and bringing the new process technology on-line as quickly as possible. Techniques that assist this are usually agile and include things like reconfiguring an existing plant using most of the assets. Another approach is the fast track where human resources are used to shorten all steps of the process. Much of the new process can be sped up because the technology is not different. Where it is different, discovery and problem solving are required. At this step it is sometimes very difficult to gain time. Another technique is to take more risk; that is, committing to things before they are understood and thus spending funds that might not be spent if a normal pace were followed. This might result in resources expenditure, either capital or technology, that may not be needed in the final manufacturing facility. No matter where a new process is required, it will have been needed yesterday. The growth of the products made by the unit will be restricted. Demand planning does not adequately predict when a unit addition will be required. It is the way that business is done and the system must be agile enough to handle the circumstance.

An agile approach to bringing a new process on-line in a timely fashion is of significant benefit to any industrial firm. It is not easily done. Having the new process technology available when the opportunity arrives is a challenge. This step, one of having new process technology available in a timely fashion, is critical to bringing the new process unit on-line when it is needed. Agility is a key factor in accomplishing this. One must look frequently at the portfolio of processes that a firm uses to manufacture its product line. The supply chain strategies must be understood. Growth rates and capacity utilization must be monitored. Each key technology must be evaluated for how well it serves its supply chain or chains. From this the process-oriented engineer can determine what technology needs developing. Resources must be committed to developing the knowledge for the new process expansion. They must drive toward discovery and learning. They must be able to capture what they learn so that it is retrievable.

The business practices selected for a supply chain can have a significant effect on the process technology and capability that must be developed. If the make-to-order, as opposed to the make-to-inventory, approach is chosen, then the capability must be built in the process area to handle the daily demands of the customers. The shift to customization of the products will require different process capability also. The process activity must respond to the desired business practices, and the right process capability must be established. The success of a firm can rely on how effectively this is carried out.

The successful development and deployment of new process technology requires that the process engineer and technologist shift agilely as the world changes unexpectedly. Responsiveness means that supply chain processes that don't experience the expected growth must be de-emphasized and those that are surging receive the appropriate attention. It also means that resources aimed at getting more production from existing assets have the same capability to shift to the product line or supply chain that is today's winner. The pulsation of the marketplace determines the degree of agility required. In many cases it goes through unexpected and explainable change. In a few cases it is predictable. Predictability decreases as global competition expands. The dimension of sourcing from another continent and another firm or supply chain adds an increased probability of an unexpected event occurring.

If the new process technology can be made available, then quick implementation can also be done in an agile fashion. One of the techniques for this is to preplan where the new capability will be installed. It may be into an existing location where space is available. It may be a replacement for other machines in a cell or process line. When defense spending declined, Motorola developed agile process lines for electronics in weapon systems. The line was very effective and was running at six sigma quality in six months from the time of initial installation. A decision was made to install a new process technology unit that brought the reliability of the unit up to the desired quality level. The new machine had been developed with the available space and function in mind and it fit into the process line. The replacement of the older process technology with the new took less than a few months. The transition was planned for and occurred smoothly. The new process unit was also needed in other supply chains, so its eventual use will be broad. The space that was vacated when the requirements were downsized went into cellular phone production use. This was a rapidly emerging technology and the timing of the two tended to be perfect. Sometimes process changes work out.

The time needed to find a space for new process technology or for machines to be implemented can be shortened by having empty or unused physical space in building or process towers. These empty spaces can become available because other units were removed, or they can be built new in anticipation of some supply chain needing expansion. This investment can reduce the time to commercialization to what is required to get the new machines or to install the process.

In a new process line or cell, much of the equipment and many of the units will not be changing. This means that designs for the whole line can exist with the unit or step being changed left as the only part of the design that will need a fast-track effort. This approach takes on a real challenge in the design and construction aspect of a new process capability. Historically, it has not been possible to repeat already completed designs at more than one location. Agility requires that this capability exist and electronic, chemical, and other companies are working to make this possible. Modular design seems to be a key to this approach. It also requires that the design be electronic so adaptation is easy. Many companies recognize that a five or more year cycle to building a new plant is not acceptable. Much change can occur during the time from the decision to build to the start-up of the facility. For example, the market or demand for the product may have changed and the plant may not be appropriate. If the time from decision to start-up could be compressed into 12 to 18 months, then the knowledge about the market would have a much better chance of being right. The shortening of the time to realization of a capability is critical in the agile world. It makes it much more probable that the unexpected can be anticipated.

Another step toward agility is to build process lines or machines so that they can be converted to other uses with little effort. As unexpected change occurs in the marketplace it is desirable that something other than what was planned or is being produced become the offering into the marketplace. If the manufacturing capability can be easily converted to the new product line, a quick response to the increased demand can occur. Developing this capability is not easy. It requires thinking creatively and answering the question "convertible to what?" The crystal ball does not easily reveal the answer. A significant amount of forethought is needed, with many of the alternatives that are developed for conversion never being used. One approach to this occurs in the rubber business where a high-volume process line would produce the main flow of the business. Other flexible batch units could be committed to the more specialty part of the offering. These batch units, if properly designed and built, could be shifted to other operations as the unexpected demand develops. Shifting from rubber to less viscous sealant material would require an upfront effort to design this capability into the units. The sealant system, if designed properly, could also shift to the rubber production. This would make the response to unexpected change in market demand shift quickly.

Besides the technical questions in the design and selection of the batch process units, other factors could restrict the ability to respond agilely. A significant one might be that the organization operates the rubber and sealant business as separate supply chains. Integration then becomes a major effort. This organizational rigidity imposes a significant restraint on the ability to respond in the desired way. Other restraints that must be considered deal with geography and raw materials. These can usually be handled if the supply chain integration can be managed. People and the response or resistance to change become significant handicaps when agility is warranted.

One method to overcome the imposed rigidity or resistance is to develop a concept of primary and secondary sourcing that requires the qualification of each product on more than one process unit. This process capability can be developed either by conversion of existing capability or building a new or more flexible process. If the qualification is in place, schedulers gain agility in how they use the equipment that is at their disposal. Developing this capability is not an easy thing to do. The vision of what is desired must be created, and process people must be put to work to develop the capability. Once in place, the supply chains or firm will be much more agile. They will be able to respond to the unexpected daily demands of the customer with make-to-order responsiveness without a large investment in equipment. Capacity will be available, and the supply chain will perfect operating in this fashion. The investments in inventory will decrease and the profitability will improve with happy customers as a result. Agility with anticipation will provide for a much improved capability.

In looking at new process technology it is necessary to consider the role of partnerships to improve the results. Working with others in the same industry was enabled by the 1984 Cooperative Research Act, which was discussed earlier. It resulted in entities like Semetech that brought the users of electronic component production machines together with the people who make the machines. The U.S. government invested significant funds to bring the industry together to make it more competitive. Industry approached the situation with caution but eventually bought in and the change was in progress. The new machines that resulted from this industry and government effort enabled U.S. industry to shift production back to home soil. The

United States became the world leader through cooperative investment in new process technology. More industries need to work together to develop manufacturing technology and then compete in the marketplace. The concept is proven and the details worked out to make it fair and supportive of the participants.

New Process Capability— Porsche

The experience of joint technology development does not exist just in the United States with government-sponsored collaboration. The Porsche company of Stuttgart, Germany, had always taken great pride in its product. The Porsche automobile, built for high performance and handcrafted to perfection, became a mainstay on the automobile racing circuit. The crafting of these high-performance cars was embodied in the concept of *technik,* or superior German technology. Although the Porsche family ran the firm for most of its history, in 1972 Ferry Porsche decided that no one in the next generation was capable of succeeding him as managing director. The firm was typically German, with the design department the elite element of the organization. It was distanced from the manufacturing element in both geography and in interaction. The firm was very hierarchical. In manufacturing, the primary workers reported to a *gruppen meister* who reported to a *meister* who reported to an *obermeister* in each of the work areas. One employee of every five was in a supervisory role. This was the atmosphere that prevailed when crisis occurred.

Porsche was building high-performance cars for very demanding customers. It was almost like the automobiles were custom produced. The manufacturing and engineering techniques could have supported that mass customization. In 1986 Porsche sold 50,000 cars with more than half in North America. The sales level was not high enough, however, to support a shift to a more cost-effective production technique.

The boom year of 1986 gave way to some horrible years when the mark strengthened against the dollar and sales tumbled. By 1992 volume tumbled to 14,000 cars, and only 4,000 were sold in North America. Change was necessary. The product design was still consistent with the high-performance image but the vehicle could not be produced at a cost consistent with what customers were willing to pay. This brought 38-year-old Wendelin Wiedeking to Porsche. He was the chairman of a parts supplier of Porsche. Previously he had managed Porsche's paint and body operation.

The change in the leadership occurred in October 1991 as a loss of $40 million was being experienced and sales were in a very steep slide. Weideking and his direct reports looked to Japan for the solution. They went on a study tour to see how production was occurring in Japan and discovered that they were far behind the Japanese in both quality and manufacturing technology. While they were studying, sales continued to drop and losses in 1992 were approaching $150 million on sales of $1.3 billion. Something needed to be done. After many trips to Japan by Porsche people at all levels, a bold step was decided upon. Weideking first brought Maasaki Imai of the Kaizen Institute into the company and developed a four-prong offensive to overcome the severe crisis.

The first prong was to reduce the company's layers from six to four by eliminating the *obermeisters* and the *gruppen meisters*. A new team structure on the plant floor was also a part of the first prong. The second step was an improvement in quality and defect elimination. It was a make-it-right-the-first-time approach. The third step was a new suggestion system aimed at rewarding workers for improvement in quality and productivity. The final step was a visual management system called *Porsche Verbesserungs Process,* or PVP, which involved setting and displaying measurable targets for cost reduction, quality, logistics, and motivation. Each *meister's* work group agreed to the targets and took responsibility.

Even with the implementation of this approach, however, progress did not occur toward the goals and more drastic action was necessary. Wiedeking contacted the Shingijutsu group, which he had met during his many visits to Japan. The leaders of this manufacturing and technology practice group, Yoshiki Iwata and Chihiro Nakao, agreed to take on the task of improving Porsche's manufacturing practices. Chihiro Nakao took charge and immediately took Weideking to the shop floor and attacked the 28 days of inventory in the engine assembly plant. Change must occur today. This was a shock and considered illegal in Germany. Change needed much study. The Japanese *kaikaku* approach focused on the mountains of inventory. Nakao knew from significant experience in both Japan and other parts of the world that inventory could be reduced. Indeed, the engine factory did bring the inventory down. When inventory was reduced from the 2.5 meter level to 1.3 meters, Nakao handed Weideking a saw and the racks for storing engine parts were sawed off and removed, a symbolic gesture that set the tempo for change at Porsche. The inventory space for engine components was further reduced to zero, and kits were used for supplying components to the engine assembler. The kits were built from just-in-time receipt of components using a *kanban* system with suppliers. The leadership of the *sensei* (teacher) had worked, and the teams were on their way to making Porsche a world-class manufacturer of high-performance automobiles. The effort continued for two additional years. The PVP teams carried the change banner and made it happen.

A major event happened in July 1994, after significant change had occurred at Porsche. A car was rolled into the assembly hall in Stuttgart. Present were Porsche's blue-coated crafts workers from the rectification area of the plant. They were there because they had nothing to do. The first defect-free car *ever* had just rolled off a Porsche assembly line or bench assembly system. Nothing needed to be fixed. The combination of Japanese *kaikaku* and *kaisen* from the *sensei* and the *technik* from the German tradition had come together to turn Porsche into a world-class company.

The focus of the Porsche effort was to take a superior product technology and develop the process technology that allowed that product to be made effectively. It was a triumph in manufacturing engineering and Japanese experience. The increasing costs in Germany were changed by enhancement of process productivity.

Changing the manufacturing process to fit the needs of a business has been a significant change around the world. There remains a significant task to improve the way companies conduct this part of their business. Competition will continue to drive firms to change the way things are done, but much room for improvement exists. It occurs in companies that have yet to grasp the concepts of lean manufacturing developed by the Japanese and in firms that need to change business practices to become more attuned to the market and the needs of the customers. Agility is necessary in making the changes and in developing the process concepts that will support the delivery of the *product* to the customer.

Summary of New Process Technology

As discussed in this chapter, the commercialization of new process technology can be done in concert with new product and market commercialization. This creates a need to make sure that all aspects of change are coordinated and managed. It becomes a very complex situation. Most process change is less complex but still requires a high degree of attention. There can be many drivers for change. It can be the inadequacy of the existing process technology as in the case of the GM stamping plant and the overall operation of Porsche. More typically, it is when new capacity is needed and the new technology replaces the existing technology. In both cases it is important that the customers continue to get the product that they have been accustomed to. Another choice is a better and different product that requires qualification in the customer's application.

When the process is changed or improved, it is important to determine whether business practices can change at the same time. The business practices determine the type of process technology that will be implemented. In GM's case, the business practice dictated reducing the investment in inventory of the parts. The capability was developed so that make-to-order could be the way to serve the customer if it were needed. Less inventory in the supply chain supported the business direction. The agility associated with changing dies in 10 minutes allowed for the competitive increase. New process technology is a critical part of moving to a more agile capability and to mass customization. The process dimension of commercialization, then, is a critical element to the success of change. It should be managed to maximize the contribution that can be gained from the commercialization of new process technology while the market and product are also considered.

The next chapter talks about the dimension of commercialization that the customer sees. The development of the new product in some cases comes from existing processes and is delivered to existing applications in the marketplace. The new product can also come from a desire to do business in a different fashion. The chapter will deal with the new product commercialization activity in an agile and mass customization fashion.

CHAPTER 10
Products

Another dimension of the new product, process, and market cube is the new product. If a new product is aimed at an existing market and an existing process, then it is not a less complex commercialization. This is indicated in Figure 10.1 where the product axis is emphasized. It deals with the existing market structure and uses the existing manufacturing process. The market must be developed and the new product aimed toward satisfying an existing need. The goal of the new product is to perform the function better than it is being satisfied by other products today. It requires the knowledge of the application and why it

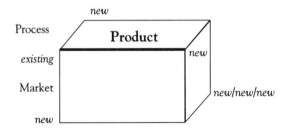

Figure 10.1 Dimensions of Commercialization

168

needs to be satisfied. This becomes the basis for the design of the next generation of product. The next product must be needed or at least perceived to be needed.

Commercialization— New Products

The desktop computer is an example of the product life cycle. Computers seem to have new capabilities at lower prices every few months or so. The ones that were initially bought just a few years ago are obsolete and can't run the robust and capable software that is available today. Thus, consumers must continually upgrade to get the full benefit of the computer capability. The time to market of a next generation in this field is very short and the product life cycle without improvement is equally short. The customer is demanding the additional capability and speed. The demand comes from the advancing software and the need for more capable computers to operate it. If the software satisfies all the needs that the customers have and slows in further development, the need to upgrade will also slow. This does not seem to be happening. Not all markets are as demanding as this one, but some very interesting needs have developed.

Quality of goods and services have become the accepted norm. At Jackson and Perkins, the service and products must be the best. Indeed, the typical gardener is very disappointed when J&P does not live up to the expectation that the company has built with its customers. If a garden plant does not grow and prosper, it is replaced or refunded without question. If the service is not up to standard and something is not shipped on time, J&P will try and make it right. Telephone communication is with people who truly are service oriented, and customers hardly ever get a busy signal. Yet the service agent takes a relaxed approach to the job. Besides the plants and other products Jackson and Perkins grows, it makes its reputation on friendly and no-hassle service.

This is the kind of service that appeals to many and specifically people who garden and grow things.

This same approach to quality exists in the catalog organizations with outdoor-type clothes. The quality is usually a step above discount or department stores and the service typically is superb. There is no question that these companies are working to have satisfied customers. This is a significant change in product and service in this industry from a number of years ago. The competition has changed for the better. Even the Japanese consumer orders from these sales outlets. A market where the product is designed and sold in America is important to the Japanese. The activity to get a new product into this sales activity requires working with the cut, dye, and sew capability in several places in the world. The garment industry exists in places like China, Mexico, Central America, and India. Levi Strauss gets some of its pants made in Thailand or Italy, but they are designed in San Francisco. Garments are really an international production activity and require an introduction of new products in that atmosphere. The lead time in the industry is typically over a year from the thread to the garment ready for retail sales. This has given some U.S.-based garment companies a chance to compete by shortening the supply chain and making it more agile. Companies like Victoria's Secret specialize in computer linking the store with the cut, dye, and sew process. Their objective is to drive the supply chain so they are making what is selling while keeping the overall inventories on the store shelf and not inside the supply chain.

Customized Product— Levi Strauss

Levi Strauss had its beginning in the gold rush days of California. The company's founder had come to the United States from Bavaria in 1847 as a young, adventuresome 18-year-old. Six years after arrival he ventured to California. He started a company that sold supplies to the gold miners and it flourished. After 20 years he coinvented a process to rivet corners on men's pants. This was the invention of the first pair of blue jeans. The jeans incorporated the rivets, accurate stitching design, and horse brand leather patches. This small beginning eventually rose to global prominence. Women's jeans were added to the product line in 1968 and Dockers were added in 1986. Today more than 100 factories are producing clothing that is sold in most parts of the world as casual work wear.

With all this lean jean clothing capability Levi Strauss introduced a new concept in tailored casual wear. This move toward mass customization involved providing a customer with a Personal Pair™ of jeans that have been formed, fitted, and customized to give the look the wearer desires. Personal fit jeans, based on the five-pocket variety, were introduced in November 1994. They are personalized to a woman's body using a special computer program to assist with the fit and measurement. It is for women who want the ultimate fit.

The personal fit jeans program works by taking four initial measurements for waist, hip, inseam, and rise. These measurements are made by a trained fit specialist and then entered into the computer, which suggests a prototype pair of jeans. The customer tries the prototype on, and fit modifications can be made based on the customer preference. She may ask for tighter, looser, shorter, longer, and so on in the four measurements. It takes two or three prototypes before a customer finds the uniquely personal fit. With the final fit the order is sent via modem to Levi Strauss in Mountain City, Tennessee. Here a dedicated team of sewing operators construct the jeans. In 10 to 15 days the jeans are available for the customer. The personal fit is stored and bar coded and other jeans can be made to the specification. The bar code number is located in each pair of jeans that are purchased. Other jeans can have additional fashion options like classic or low rise style, different denim finishes, and tapered or boot-type leg openings. All this customization can be had for $65 per pair.

Levi Strauss's move to mass customization is an example of a company recognizing a need and providing the product to fill that need. Jeans developed from a rugged work pant for the gold miners to fashion-oriented casual clothes. The company developed from make-to-inventory to make-to-order, introducing custom clothes to make a personal fit possible. The new product line required a method for measurements and a connection between the customer or store and the manufacturing facility. It sets a standard for customer satisfaction that others will follow.

Other products are also made to satisfy a need. Likewise, they are aimed at existing markets and use existing manufacturing capability to produce. A new formulation for a window sealant in high-rise buildings can be formulated to perform the demanding tasks of not only sealing out the weather but also holding the glass on the building. Once proven and demonstrated on the design test stands by imploding and exploding the glass panels, the new sealant product can be produced in equipment that exists. The product is customized to the architect's specification for the high-rise building that is being glazed and sealed. In doing this, it is important that the process and the product are capable; that is, the process must be able to produce consistent material in a reproducible fashion. All the windows must stay on the building and be weather tight. The process needs to perform to a six sigma level of quality. This high quality is necessary to ensure that the product performs as expected. The application is very demanding, and the consequences of either a seal failure or an adhesion failure are significant. A failure can take a number of forms. It might be a discoloration due to aging. It could be a break in the ability to exclude the weather from the building. It could be a dirt pickup of the sealing materials. It could also be a loss of adhesion and glass panels falling off the high-rise building. The falling glass is not an acceptable situation.

To create six sigma quality, the process must be capable, and the product must be well developed. The system must be agile because the architect wants not only the performance just described but also a desired color and texture. Color matching must be part of the agile capability. Pigmenting to the desired color must be done at the point of packaging to provide for the most responsive system.

New products have unique requirements. They must perform better than materials or products introduced into the market a cycle earlier. Customers demand continued product and service capability improvement. To be able to respond, the process of product or service introduction must be agile. The marketplace dictates the need. The product development people develop products that meet those needs.

The process of design and testing needs to occur rapidly so that the commercialization process is as speedy as possible.

Commercialization Process— New Products

The new product commercialization process must be agile to be successful. The continuous change in customers' needs requires a rapid response by a firm to be competitive. Agile practices for new product development must be developed and become a part of the firm's tools. To develop this capability, the existing process development methods must be understood. Each step of the process should be analyzed to determine the amount of time that each step is taking and why the process is as long as it is. The process also must be assessed for value-added time as a part of the total development time. The goal will be to eliminate idle time and increase value-added time as a percentage of the total. With the understanding of the existing process, a more streamlined process can be developed. One of the key focuses must be the activity associated with the initiation of a new product development program.

The start of new product development requires a special effort to be done right and not stifle creativity. Customer needs must be developed, and focus must be on market and customer-oriented techniques to start the development cycle. Assessments must involve the sales force who understands customers and their needs. The idea for the new product might come from the R&D efforts internal to the firm. These ideas must be skillfully brought to the customer for assessment without promising that the product will be available. The sales force can also bring knowledge about the product offering or expected offering of competitors. Competitors' activities in the new product area are useful in determining what the application or market will be like in the future. The R&D people need to gain insight from the sales force to adapt their thinking and technology to make it consistent with the customer's need and situation.

Other product ideas can come from looking at the existing product line and determining where it might not fully meet the market or customer requirements. This may start as a fix for the product line but could grow into a new offering. Using the others in the firm who deal with either the production of the existing product line or the customer may provide information on deficiencies. For example, the production people can describe problems that affect the ease of manufacture. They can give insight into the complexity of the production process and the places where new technology will make the product line more effective. The people who interact with the customer, like order entry or quality people, probably understand where the product line can be improved. These sources should be part of the decision to commit to a new product development program.

Another source can come from outside the company or firm, through benchmarking with others and thus gaining a feel for where an unfilled need might exist. Other outside sources might be inventors, suppliers, or vendors, or the customers themselves. Each could describe a need that may be different than what the existing product line satisfies but that might align with a technology that the firm has. The technology might be used in a different part of a firm's product line, or it might be a technology that is in an early stage of development. Applying the technology to a need requires, in many cases, trying to fit in the ideas that others have. With the source of the product ideas coming from many places, it is important to make the decision to do the product development as quickly as possible. The quality of the decision must also be at the highest level possible. Once the commitment is made to go in a particular direction, it is important to agilely steer the effort to success.

As the concept of the product gains support from the market and customers, an effort is essential to match the new product design with what is needed to manufacture the product. Design for manufacturing becomes a significant part of being able to produce the product in an effective fashion. The more work that is done on manufacturability, the higher the level of agile response to changing customer needs can be. Agility or custom manufacture requires that the new product or products be produced from standardized parts or materials. It requires that these be kept to a minimum. A modular design should be used and the characteristic or identity of the product should be developed as late in the manufacturing process as possible. The parts or materials should be able to be combined to make the products perform differently. The new products should be able to be produced on machines that are standardized for all the products. These machines or the manufacturing process must be effectively designed and laid out, and easily operated. Setup or conversion times from one product to another must be minimized. Quality must be integrated into the new products with defects minimized. New product commercialization success requires that products meet customer needs and be produced in an agile and effective fashion. This is much more important as the new product concepts are introduced into the marketplace where the real needs may require the new products to be adapted. The activity associated with new product commercialization is usually a disciplined process with well-defined steps and a streamlined and efficient work process. Speed of commercialization is a key component of new product success.

Most companies that introduce new products have a work process to guide the activity. This process usually has specific milestones to define product development. These milestones are sequenced to an initial climax of commitment to make and sell. Different industries treat product commercialization differently by dealing with the risk assessment for the new product and what risk is acceptable. Manufacturing would like to see enough production occur before the commitment to demonstrate that the process and product are statistically capable, ideally a distance to the nearest specification of greater than three or more sigma. This cannot always be accommodated, however, and the product may be sold with just a small production experience. A product that the customers like enthusiastically will get lots of manufacturing attention and will move down learning curves rapidly. This will minimize risk. A product that does not sell well, and that only one or two potential customers like, presents a problem. Moreover, a firm never gets good at producing these types of products, so they are always a problem. The process to develop and commercialize a product tends to stumble at the stage where it should be dropped from the line. Firms with a lot of new product introductions must have a commercialization process that works for all circumstances.

Figure 10.2 describes the new product development process and how it impacts a firm's supply chain. The process starts with gathering information on needs in the marketplace from both internal and external sources. This results in a formal decision to pursue developing a new product to satisfy the perceived need. This starts the process of better defining the market needs and the scope of the product. The market size and potential customers are also determined. After the scope and the market are defined, the process of determining the technology that will be used in the product is brought into focus. This is an assessment process that the New Business Product Project Manager and his or her team take very seriously. The technology may need more development, but it cannot be just a gleam in a researcher's eye. If the new technology is critical to the competitiveness of the new product, a commitment can be made to develop the new technology to a useful form. Backup alternatives are necessary in case the new technology does not come along in the right time frame.

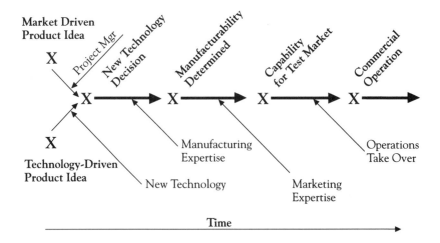

Figure 10.2 New Product Commercialization Process

With the scope of the product defined and the product development process committed to and under way, the new product now needs a process to produce it. This involves adapting the product so that it is manufacturable. The determination of the manufacturability and the manufacturing process is not a simple step and should require as much resource effort as developing the product itself. This step, combined with market acceptance of the product, will determine the success of the commercialization.

The new product can be offered to the market once the decision to make and to sell is made. This requires acceptance by key customers and the manufacturing system. With that complete, the new product becomes a part of the supply chain or firm's product line. The speed with which the whole process is conducted is critical to enhancement of the competitiveness of the enterprise. The effectiveness of the process and the degree of knowledge development determine how agile the product offering can be. The more knowledge developed about the application, the parameters of the product, and the manufacturing process, the easier it will be to respond to unexpected needs of the market or to develop a custom manufacturing capability. Many firms have developed or adapted the basic workflow process shown in Figure 10.2 to meet their specific needs.

Kodak has developed a unique process that meets its new product development needs. The process starts with research people spending a significant amount of effort to determine whether a new science or technology is well enough developed to be a part of an upcoming new product. This exercise involves critical evaluations of the benefits versus the risks of premature introduction. It takes into account that a new technology could be brought along as a product is being developed but that at some point the design must be fixed. This technology interaction climaxes with the selection of the technology that will be used and the identification of those technologies that will need to be developed further. At this transition point a new product is taken over by the operational business unit, which manages the product to commercial introduction and growth.

The operational business unit must decide what the final product features are—what it will look like and how it will be designed so it can be made. The work process is very much an interactive one with prototypes being built and tested. The team works to make sure that the new product will fit a perceived need in the marketplace. They also work to ensure it looks and feels good in the eye of the customer or user. They work to see that it can be produced within the capability of Kodak. This parallel step process eventually brings in the senior operational people, who have a final look at the manufacturability of the product. This is a very critical encounter, one that takes many sessions to iron out the concepts of making the product right and at a cost desired. It is done with interaction of the team, but the manufacturing experts carry significant clout. They have the scar tissue associated with other new products that were not worked over for manufacturability. When everyone is satisfied, the product is launched and the marketplace will decide whether it will be a success. With Kodak's broad product offering, this process is used on everything from cameras to film and from projectors to x-ray imagers.

Commercialization — Speed to Market

Agility in the new product introduction process is aimed at speeding up the introduction and ensuring that the product meets the customer's needs. Agile techniques are necessary to do both.

The point of developing various aspects of the product simultaneously is to introduce the new product as quickly as possible. The automobile industry has been working on the workflow processes of designing new models to shorten time to market from what had been five years to under two. The Japanese automakers, who have led this effort, have been successful in having the right model available to the market at the right time. U.S. automakers have also made significant improvements in the speed to market. The parallel path design that is aimed at shortening the commercialization process requires a disciplined but agile practice. As design develops, various parts must be modified for the sake of the whole design. The tools must exist to quickly adapt the design to fit the concept of the new model and the components that will be assembled. The overall impact will be that the needs of the marketplace will be served with models that better fit the customer requirements. This shortening of the time to market is occurring in many parts of industry. Its importance cannot be overstated.

When key leaders of the camera industry were on the verge of changing the technology by incorporating electronics into the camera and the film, they also were replacing the film completely and going to the electronic camera. The leaders got together and worked to standardize the various components of the technology so that things like film and printing could be interchangeable. It was a cooperation that recognized the importance to each firm of having something that the customer or user can consider as interchangeable. Memories of Sony's beta format haunted the industry leaders, and they felt it was prudent to get together before the commitments to one format or another were made. This would shorten the time for new camera technology to get into the market and ensure the success if the overall concept of the new cameras was what the customer needed. Each producer then went on its way in competing for the business. Kodak chose to have its first version of the electronic film camera produced in Monterey, Mexico, in a streamlined and modern plant.

Other times, a product does not perform as the customer would like it to. In these cases, the enterprise or firm must be able to respond agilely and with speed to modify a product to meet the expectation of the market. This ability to agilely and speedily adapt the product to customer needs is a capability that must be developed. It does not just happen; it happens because the unexpected is anticipated and preparations are made to handle the circumstance.

The development of a new product requires a large amount of interaction among people on the team, which is usually controlled by a work process with various steps driving to the end results. If all this works right, then the product will meet the user's need. If not, then one must provide a fix in an agile fashion. It might be the modification of a component that is incorporated into future production. It might be a retrofit of the released new products to make them more capable. It might mean a quick introduction of a new and improved version or model that competes with or eliminates the original. In each case, the team must be prepared. The capability to assess the performance of the product in the market and to make quick modifications must exist. It must have been planned for and predetermined that it will be required. It should not be a red alert scramble but a planned event that works effectively and quickly. The success of the firm is at stake so remedy must be swift.

Summary— New Product Capability

This chapter has described the concept of commercializing new products into the marketplace. It dealt with the definition of the new product and the sources of information that define the need for the product. It also discussed fitting the new product to the business practice that is desired for the product line. Customized products were discussed, with Levi Strauss's Personal Fit™ jeans for women and a sealant and adhesive for high-rise glass buildings being examples. The concept of a process for new product commercialization being important was introduced and a proposed process was defined. The importance of time and speed to market also was emphasized. Commercializations usually do not involve just one element or dimension of the new product, process, and market cube. The Dimensions of Commercialization cube includes both the process and market in addition to the product. These three dimensions show the complexity that can be encountered when dealing with the addition of capability to an enterprise. The discussion has focused on new process and product of the cube.

The next chapter will deal with the new market. This is a critical dimension because it is where the customer resides. It tends also be a place where not enough disciplined attention is given to defining the market needs. This will be followed by a chapter that discusses people and relationships within the commercial enterprise. Things like the strength of relationships, the human organization, the knowledge worker, and the importance of the individual will all be discussed.

CHAPTER 11
Markets

Developing business that a firm is not currently in requires the assessment of the fit that the firm's products have with the applications within that market. The market can be a segment of a market the firm has already been doing business with or it can be an expansion into a new region where the product line will likely sell. The new market may be an extension of where business is done today or it might be taking a product line national or even global. Each of these are ventures into the unknown. This situation is described on the Dimensions of Commercialization cube as the vertical axis in Figure 11.1.

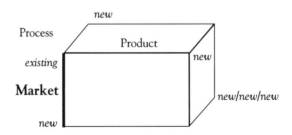

Figure 11.1 Dimensions of Commercialization

This chapter will deal with new application of existing products in the markets currently being served and into geographically expanding markets. It will also deal with expanding existing products and existing applications into new regional, area, or global markets. Both new and old applications of existing product provide a significant challenge when they are looked at as expanding into global markets.

New Applications

New applications and opportunities are available to an enterprise in a continuous stream. A very important part of the new business commercialization process is determining which opportunity will be selected. This is not always a straightforward decision. The first reaction should be to determine whether something from the existing product line could fit the need. The second reaction might be to see whether an existing product can be slightly modified or adapted for the application. Next might be to see whether a product could be developed that can be produced on the existing equipment. This process becomes really complex if the application is in a new market where the firm has not done business and that it does not understand. The knowledge that is needed in the new market is sometimes hard to get and can be expensive to learn by doing.

A new application is a potential source of profitable growth. The investment has been made in the existing product and process technology. That technology has been used to satisfy a market need, product volume has been built, and the manufacturing capability has been established. From the product and process perspective a new application is a freebie. It is not a freebie from the market perspective. Defining and satisfying the needs of a new application require an organized and significant effort. It might even mean the establishment of a new capability to get to the customers that is different from the existing business.

Market development people must look for applications that would be good business for the enterprise. Some of the applications cannot be satisfied with existing products. These need to be evaluated to determine if the opportunity is large or profitable enough to justify the new product development activity. The applications that look like they can be satisfied with existing products need further evaluation and screening to determine if the effort should be expended to pursue the business.

The evaluation and screening must include the determination of whether the existing product not only works in the application but also whether a true advantage exists. This advantage will be required to shift the application from today's solution to the firm's product. This advantage, if it does not exist, makes the firm's product a "me too" product, and the only way to compete may be on price. That type of business is probably not what the firm is looking for. If the advantage does exist, then the initial screen is passed and further information is needed. The size of the potential market needs to be assessed with more certainty than was done in the preliminary screening. The types and variety of customers must be determined. The pathways to these customers require definition. The pathways must include logistic routes for the products, information routes for orders and other communications with potential customers, and the method for cash to flow for payment. Each of these pathways is critical for building the right capability to serve the customer and to determine whether the firm can be successful in entering this market.

The logistic route to the market must be established with the shipping methods, the type of interfaces with the shipping method, and the inventory strategy requiring assessment and commitment to a procedure. The shipping method can range from air freight to parcel post to trucking. Many options exist, but one is probably right for each customer. The interfaces between the method of shipping and the firm as well as the customer must be determined. In some cases the customer will give the producer and the shipping company a 30-minute window that must be hit, which requires significant coordination and the building of special relationships to make it happen effectively. Other interfaces are simpler to establish and require less precision. These interfaces must be satisfactory for each customer and may require developing agile relationships.

The inventory strategy should fit the new applications. The key is to minimize the firm's investment. The ideal, if the product is an existing one, is to make to order and ship in a just-in-time fashion from a continuous flow manufacturing and delivery system. This requires an agile approach to these new applications, and the existing product sales must have established this type of capability. Make-to-inventory and delivery from the warehouse would also be an alternative. In both cases the customer must be effectively served and the investment in inventory must be minimized. This establishes the pathway that will get the product to the customer.

Another pathway is that of information. The method to establish orders and demand is a critical communication to establish. In some cases the scheduler with the producer of the product can be in contact with the customer and can enter orders and determine what the demand for the product will be in the next week or so. The more usual approach is through an order entry capability with demand being determined by contact with the customer's purchasing department. If vendor-managed inventory becomes one of the advantages of the firm's product over the others, then a monitoring system needs to be established to make this effective. Other links are important also. These include receipt of goods, invoicing, product performance or quality situations, and changes in the product that might affect the performance in the new application.

The final pathway is the transfer of cash to pay for the product received. A link needs to be established that makes the payment. It is also important to have a communication route to determine the terms and conditions of the product. This may be a different link from the one established in the pathway for product or information. Establishing terms and condition of sale and purchase is a strategic negotiation and can be separated from the more operational pathways.

The new application for an existing product can be thought of as something simple to carry out. In many cases it can be that but the need to be more competitive in delivering a new solution to a need may dictate that more consideration be given to the pathways to and from the customer as a competitive advantage. This means that significant effort may be needed that is beyond establishing that the product will work in the application and the price makes it a good business opportunity. Agility is required to develop the right kind of relationship for all customers. Expanding the market for existing products makes good business sense. New markets will add to the growth of the firms and will expand the use of developed technology and existing capability.

The new market where the new application resides has unique features that must be determined. These features must be understood and built into the plans discussed in the preceding paragraphs. Questions to ask are those such as the following: What is the new market's commercialization cycle? Is it quick? Does it go through extensive testing and is there an assortment of candidate solutions for the customer's defined need? Where is the decision made on the solution? Is it at your customer's enterprise, or is at your customer's customer?

This list continues relative to getting into a new market commercialization cycle. There are also factors that affect the way business is done in the market that might not be understood: Are all deliveries just-in-time? What are the terms of payment? How aggressive will the price competition be? Where are the buying decisions made? Does a qualification from the engineering part of the firm make your product offering something that will result in business? Does the qualification just allow you to compete in the procurement activity? Are you being used to extract a better deal from others, and will you get business even if you meet all requirements?

Since an extensive amount of knowledge is not available when looking at opportunities in new markets, circumstance and a significant future opportunity must be anticipated to invest in learning how to do business in a new market. If it is a one-shot opportunity with no other use of the market knowledge, there may be other ways to get to the application. Finding a firm that knows the market and partnering with it to gain entrance may be a way. It involves giving up some of the revenue and profit, but it eliminates the need to learn the customer and the application in detail. A distributor or channel partner may be ideal to serve this need. It could also be a business that satisfies almost the same need and would like to fill out its line with your product. This could either be under your label or be custom packaged under that firm's logo. Clearly, a wide array of options exists to get into the new market and begin a learning process.

Using a partner is difficult when trying to determine what product will be offered. The choices are to teach what the product can do for the partner. Partners can then work with the customer to fit the application. If the capability does not exist within the partner to provide that service, then it may be necessary to work with the customer directly, with the partner in an advisory role.

Another difficulty is in determining how the revenue and profit will be divided. In many cases, greed from either player prevents a partnership from forming. A company may not have an appreciation for the knowledge about the market that the partner has and does not value it highly. Likewise the partner may feel that a lot of pioneering with the products would need to be done to learn how they would work and how the partner would approach selling them. The uncertainty on both parties' part can result in failure to tie the knot and start doing business. This gets strained further if the product does not do the job or if the sales don't start coming in.

Taking a company's products into new markets is not an easy task. If it were, we would see more expansion of this type of business. Many reasons can exist for the lack of demonstrated success. A primary one is the lack of knowledge of how to do business in the market. Another is what the new customer expects in bringing your product into his or her company. Some markets and customers are staffed with people who can incorporate your offering into their designs. Others require working with outside design and manufacturing to integrate your offering into their product line. The security that one gets from knowing the market can be very appealing and the risk of entrance into a new market overwhelming.

New application in new markets continues to be a source of good business growth. It is a prime method for expanding the sales of a product or product line. However, it does have many challenges and must not be considered as something very simple to do. Effort consistent with the eventual benefit must be expended. Agility must be applied to developing both new applications for existing products and ways for customers to be serviced. Another source of existing product growth is through geographic expansion, either to another region of the United States or in a global fashion. Both provide new challenges for the enterprise.

Growth through Regional Market Expansion

Taking an existing product or product line from a regional focus to a national market is a significant challenge. It is handicapped by the lack of knowledge about the region targeted for expansion. Therefore, the market or supply chain team must develop the necessary knowledge about the new markets. Most of the learning will be the same as when finding new applications or customers for the existing products within the region where business is currently being done. This has been discussed in the preceding paragraphs in some detail and will not be repeated here. However, some differences do exist.

The product applications are proven, and existing business can be used as an example of where the products work and what the benefits are. It is a foot into the new customer's door and will enable getting a chance to show what the product and the firm can do. In regional expansion there still must be advantages to the products being offered. To be successful, they must displace the existing products that the customer is buying. Performance or other advantages must be good enough to overcome the increased distances from the point of manufacture. The logistic chain will require development to successfully and effectively get the product to new customers. It may even require that points of manufacture get developed in the region where the expansion is occurring.

Each large potential customer must be contacted and evaluated relative to the role that the customer can play in the strategy to enter the region. If the evaluation is positive, then a relationship must be established that results in the new customer playing a role, and possibly a special one, to help in entering the new region. It is important with many product lines to develop the needed volume in the new region that enables the investment in resources to support the presence of the product line. Each type of product that is expanding to a new region must be treated individually. Some will require a manufacturing presence. Some can be handled by catalog mailings. Others will require a special relationship with potential large customers. It is important that the right strategy be developed and executed.

Growth through Global Market Expansion

In the case of expanding a market into a new region in the United States or North America, the cultures are diverse but operate with a common thread. They tend to communicate in English and respond to the market as Americans. There are more similarities than differences. This is not true when one looks at the world as one's market. It is filled with different cultures and thinking. Various regions also have different levels of competition. Levels of economic development are different, and countries are growing at different rates. Asia is the fastest-growing region, and many businesspeople are targeting it for product expansion.

Asia is not one market but a series of country markets that are very different. Japan, South Korea, China, Taiwan, Malaysia, Singapore, Thailand, Philippines, and Indonesia are all in various stage of economic growth. Relative to other countries, Japan's growth has flattened. South Korea has been an economic dynamo as of late. China has been growing and acquiring technology very rapidly. Others are also industrializing quickly. Each has a unique situation, and expanding into these markets will take adaptation to the local situation. It will require agility on the part of the global manager and his or her team. Circumstances will arise that will require flexibility and knowledge to reach the business opportunity. Language will be a problem as well as understanding local cultures. A real challenge will be to establish enough presence so that local people can be part of the team that develops the business.

Some companies establish their presence by building manufacturing facilities that focus on both an export market and the local market. The people they hire will be trained to manage and operate the manufacturing facility. This will form a foundation for the selling of the product line into the local economy. In some countries, it is possible to sell product that is imported, which is used to fill out the product line. In others, the import of products is discouraged through tariffs. Just what strategy fits each country or market depends on the situation and the stage of market development. There may be as many strategies as there are markets. In someplace like China three or four different markets may exist that require slightly different treatment.

In most of the Asian countries mentioned, a large percentage of the economies are influenced by resident Chinese. John Naisbitt, the author of *Megatrends Asia*, calls this phenomenon the Chinese Overseas Network. This ethnic Chinese network is used to increase the opportunities of industrial and business expansion for the people of Chinese ancestry. The network is not something easily tapped into by others, and it requires a special understanding and relationship, one that is uniquely Chinese.

Asia is a part of the global economy that many companies have focused on to expand their markets. Many companies have gotten very good at this, but not without experience and effort. Some 30 years ago I participated in expanding into Japan from business that Dow Corning had in Europe and the United States. Initial effort was with imported materials and products and with technology cooperation and licensing. This advanced to manufacturing with both a wholly owned finishing operation and a basic and finishing joint venture. This enabled Dow Corning to be a significant part of the Japanese market. Today the joint venture and the finishing operations are run by local people with a minimum of involvement by foreigners. The primary interactions are strategic and technology exchange. Technology flows in both directions; thus, the technology developed for product and applications in Japan is used in other countries throughout the world.

This success in Japan is the model for other parts of Asia. The difference may be that joint ventures will not be a part of the strategy. Hiring good local people will be what establishes the presence. This has worked for South Korea and Taiwan and is being planned for China. In China, initial hiring and market development are occurring, and a finishing facility is operating in the Shanghai region of the country. The strategy will be to continue to expand the market and gain as much share as possible. It is the desire to be a dominant force in Asia. As Japan loses its economic superiority, the rest of Asia will become a global force and a significant part of the global economy. The strategy does include being agile and doing what is necessary in different countries of Asia.

Establishing a presence in places like Mexico or Central America requires understanding the culture and the way to do business. It is not much different than what is going on in Asia, but the growth rate is not as high and the economies do not seem to be emerging with the same intensity. To expand the product offering to this region of the world will require different approaches that are unique to the cultures that exist in these countries. Neither language nor local people will be the same concern as in regions more distant from the United States. There is enough influence from the U.S. culture and opportunity for interactions between countries that this will not hinder the development of a presence.

In Europe, where the economies are developed, the conditions of doing business are well understood, and one must fit the norms that exist in the various countries. Introduction of products to these markets and expanding into Europe can be looked at and benchmarked with others that have done it recently. The countries of Eastern Europe present a specific challenge with very little growth and a surprising absence of wealth. Indeed, some of the problems deal with getting a fair value for the products delivered.

Each part of the global economy will require special effort as product expansion is assessed. There are similarities in what needs to be done, and each country will require an agile approach to the tactics that will be used. It is an opportunity for expanding the growth of a firm and taking advantage of the technology and capability that have been developed.

Market in the Commercialization of New Products

The development of new markets not only focuses on taking existing products into new applications or segments, expansion into regional markets, and global market expansion. It is also required in the development of new business as it relates to new products and new processes or supply chains. The concerns in the market are the same as described in the first part of this chapter, but added complexity occurs with two additional needs for development. First, the new product has no proven track record and that must be developed with applications from either new or existing customers. Second, complexity is added if the manufacturing process for the product is new also. New products, markets, and processes represent a real challenge for firms. If the technology is also new, then the most difficult new commercialization expands the challenges even further. This situation has been discussed in Chapter 8 but further comment is appropriate here.

The new market for a firm's product requires that much learning and decision making occur. If it is to be successful, a strong foundation or infrastructure needs to be developed in addition to relationships with the new customers. If the driver is also a new product produced in a new process that will go into this new market, the tasks start to approach the impossible. When Motorola encountered this situation for cellular phones, a special approach was used. The cellular phone development entity within Motorola was given special treatment and reported to the president. The same thing happened at Corning when it was developing optical fiber for communication systems. The program was treated in a special fashion and reported to the top of the organization. In both cases, the task was looked at as too big for the existing structure.

These two cases, and there are many more, illustrate what it takes to bring all the needed new technology forward and into a market to get it established. In each case the intention was to be in a global market with lots of needed infrastructure to support it. In the case of Motorola, the company needed towers on the ground and satellites in space. It needed stores to sell the phones in most parts of the world and an organization to run the networks. It all had to come together. It was not an easy task without risk. Motorola took that risk and has been successful in developing this significant change in technology. A cellular phone is the device of the day, and everyone needs to have one. The copper wire or glass fiber of a conventional system will not be installed where it does not exist currently. The concept of this communication device will change and will involve using the airwaves to talk and send data. Exactly what role it will play is currently being defined, but it will change the world. Optical fiber did the same in changing the amount of information that can be transmitted on a single cable. It allows for expansion of the amount of information and conversation that can be communicated without a significant change in the historic concept. Copper was replaced with fiber and capacity expanded.

In both of the preceding cases, the market needed to be developed while the technology was emerging. Both the product and the process to produce it were needed. To be successful, organizational structure needed to change so that the higher level of risk could be deemed acceptable. At lower levels in the organizations, the bold and creative projects could not get the right type of nurturing. It took a commitment from the top. As change occurs and commercialization activity is evaluated, it is important to consider the magnitude of the change. This probably varies by firm, but creating mass customization where it does not exist is that type of change: It involves the market, the design of the product, and the process or supply chain. Each must extend or create new technology so that the proper role is developed and makes the whole commercialization successful. To accomplish that, agile tools are needed to support the shift.

Summary of New Market Capability

This chapter has looked at bringing existing products into new applications with either existing or new customers. Many activities are needed to be successful. They vary with the situation, but many things must be considered. The change process starts with assessing whether the existing products can perform in the new applications and whether that performance will provide an advantage to the application or customer. If the answer to this is yes, then the hard work starts of building relationships and infrastructure to get the product successfully to the customer.

That same infrastructure and new customer relationships are needed when a firm decides to expand to another region or go national with existing products. Working in the marketplace will require a well-developed and agile plan of attack. It is a learning experience, and as knowledge is gained, the approaches must change to be successful.

Global expansion adds a number of dimensions to regional expansion. Being able to do business in a new culture with a different language requires some special infrastructure—local people who can become part of the effort. This is the most challenging part of the effort and many people with different language skills and from different countries will need to be brought on board and trained. These are not just the people who will interface with the potential customers but also those who will be involved in building the infrastructure to do business. Global expansion, like regional, requires that the plans be agile and that frequent reassessments are made.

Market activity plays a key role in the more risk-oriented activity where a new technology is being brought forth that requires a new product and new process or supply chain and is aimed at new applications. Special considerations and plans are appropriate for this type of activity if it is to be successful. Commitment and understanding are needed for each step of this complex process. It may be that the only people capable of managing the risk are at the top levels of the organization. A shift to mass customization where it has not existed before may require this type of management. Agility will be required in both the process to build the capability and the activities as implementation occurs.

Attention to the market when expansion or new commercialization is the direction of the firm is essential to success—something that requires very intensive and creative effort if the effort is going to yield the desired results. The market effort should not be understated or considered a part-time effort.

The next chapter, in the fourth and final section of the book, will deal with people and their importance to the organization. Relationships become the foundation of the enterprise and the basis of all that is done. Cooperation is needed to compete and it must be nurtured. The type and strength of different relationships must be understood. They must be matched with the desired activity between people or between firms. They need to be preplanned, not just treated casually.

The human organization becomes the tool with which the organization operates. The treatment of people will influence a business's results more than any activity. People will make a difference, so it is important to understand them and the organizational approach that will make them and the firm successful.

PART 4

Pathways to Agility— The Journey

CHAPTER 12
Agile Relationships

None of what has been discussed in this book is possible without skilled, motivated, agile, and effective people. The people at all parts of the supply chain, from customer to vendor, are the topic of this chapter. People and their relationships do make a difference. They are a paramount consideration and need the full attention of all members of the team from leader to implementer.

This chapter will develop the concept of agile relationships. It will deal with specific levels of various interactions and what they mean in different cultures. The chapter will show how cooperation and partnerships enhance a firm's competitiveness. It will discuss the strength of different human interactions from the relationships between team members to spot purchases by customers. It will deal with the human organization as an essential part of the successful enterprise. It will describe how people make the difference when striving for an effective enterprise.

The nature of people as they fit into the various societies around the world can be an important consideration. Because cultures differ, acceptable actions in one part of the world may be taboo in others. However, some characteristics seem to prevail in most parts of the globe. All societies place a high value on a person's character; that is, they expect people to have integrity and to focus on doing what is right. People are expected to have courage and perseverance. Hard work, whether it be physical toil or mental challenges, is also expected of individuals. However, cultures are often perceived differently because of variations in traits and customs.

Places like Korea or Japan exemplify the ethics of hard work and being valued by the employer. Playing fair in competition is an expectation and fairness is interpreted as following the unwritten rules of the game. Intense competition is possible and respected as long as those rules are followed. Discipline and perseverance go hand and hand. Keeping focused on the direction and making continued progress to the goal are needed for an individual to meet the expectations of others. Following through on the promises that are made is expected. People are expected to take and keep the responsibilities that come their way. Responsibilities are not assumed and then cast off at the convenience of the individual. Finally, the individual must see the positive in others and have respect for them. This respect for others must be genuine and will be felt by others as that. It is a foundation for building strong human interactions.

The next part of this chapter will deal with the importance of focused cooperating and partnering to enhance competitiveness. This will be followed by a discussion of the human organization and why people will make the difference in striving for an effective enterprise.

Cooperation to Compete

Earlier parts of this book talked about operational agility and developing new capabilities. In each of these areas, good relationships are necessary for the activity to be successful. The ongoing relationships along the supply chain take different forms. They can be the cooperation born out of necessity as individuals are brought together to work as a team. They can be relationships that develop as one extends the supply chain to the customer or vendor. They can be developed with service providers who assist the supply chain. In all cases, it is important to develop relationships in order to be an effective part of the activities of the supply chain. The relationship operates on a personal level with a focus on the personalities involved and can also focus on the business relationship that is desired and the relationship that exists between the enterprises involved. In most cases there is some of each kind of relationship in productive cooperation.

As the world changes, relationships will need to adjust in the same fashion. In the past, relationships took time to form and then had permanency. In the new global world, the process of building relationships will require more speed. Good relationships are still required for success but will need to be built in a more productive fashion. In many cases, relationships need to be fashioned or developed quickly but can be discontinued just as quickly. The speed of change makes it critical to understand that relationships are important and that the right kind must be developed to support the expected business activity. The development of the relationship must be looked at as an important step in reaching a business cooperation.

In this changing world it may be nessecary to develop relationships with entrepreneurs who have their bags packed and are ready to flee a failing situation and leap to an opportunity. The emphasis will be on the speed with which opportunities are shaped and then discontinued when they have no further importance. U.S. businesspeople, who have prided themselves on solid long-lasting relationships, will be challenged and will need to acquire the ability to move with speed. Today's businesspeople will need to take measured risk now to enable competitiveness in the future. Those risks will strain old relationships as they change and stress the new ones as they are formed in a speedy fashion. Businesses must look at a global market and market opportunities. As business is executed, either with customer or supplier, the focus must be on the people where interfaces are developed. Making them strong and effective will require special attention.

An example of the need for strong and effective relationships comes from a company located in the U.S. Pacific Northwest. Boeing represents global leadership in the airplane industry. It excels at the production and maintainance of complex systems that move people through the air safely. Boeing has determined its core competencies in the airplane business. They outsource the manufacture of various components but continue as the industry leader by being the best at the integration of the aircraft components and the design of critical key or core elements. They are skilled at getting others to make components that come together to become an airplane. They specialize in bringing all the parts together and focusing their expertise on the design and specification development. This is an example of an enterprise that relies on others to do much of the work in the manufacture of an aircraft but retains the people skilled in the corporation's core competency.

To achieve this ability to farm out and then integrate, Boeing needed to develop many relationships with others and cooperate to make it all work. The precision of the integrated activity can be seen at the aircraft assembly operation in Everett, Washington. Components arrive and are used in a just-in-time fashion from all over the world. They are incorporated into the aircraft at the right time. Eli Whitney, the inventor of interchangeable parts, would be amazed at the speed and precision of the operation. There is no significant inventory of components. Parts are produced and delivered as needed in the assembly process. Each aircraft is on a schedule, and the suppliers are producing for a particular model and plane. Nothing is produced for general inventory, yet the assembly occurs with precision.

Boeing has a network of trusted and efficient suppliers with whom it works at building trusting, solid relationships. The activity is all preplanned, and the role of each supplier is worked out and defined. The company's core competency may be building aircraft but it also has one in building effective relationships. Partnering is a way of doing business, and Boeing uses it not only with its suppliers but also with future customers. For example, Boeing involved all its potential customers in the design and specification of its latest model, the Boeing 777. Intensive interactions shaped the design on every detail. The design team focused on everything from passenger comfort, pilot ergonomics, and cockpit layout to maintainability. It was a significant new design with the development of new relationships from suppliers to customers—relationships that will keep Boeing as the leader in this industry.

Relationships in a Supply Chain

Within the supply chain relationships are also paramount. No matter where in the world the supply chain operates, to execute effectively, it will be necessary to address the people involved. They must be people who represent the diversity of the local economy but who also have global capability. There will be people from each locale who are best suited to operate locally, with some interfaces extending into the global business community. Others will thrive in a global commercial activity and be able to do business anywhere. From each local setting it is important to identify those people who will become global businesspeople and provide them the opportunity to develop that skill. The successful global businessperson will have the talent to develop productive relationships in the many cultures of the world.

The supply chain will operate with teams of people who need to work together and across teams to be successful. In every team, circumstances will exist where harmony is not easy to develop. It may be due to any number of things, from competition between individuals to private conflicts of personalities. Whatever the cause, change needs to occur so that teamwork is established. The team must assess the talent needed to be effective and make sure that it is present or being developed. As in any chain, a weak link, relative to a skill availability, will make it very hard to succeed. The teams and the integration across the supply chain must work to provide a well-developed capability that has the tools to perform effectively. People must rise to the need and develop new skills or seek out people with the needed capability. The development of cooperation between capable people requires the harmony of the needed skills.

In a supply chain, the relationships between customers and suppliers warrant a special focus. Just what needs to be developed varies from situation to situation. Some vendors of core raw materials require a relationship that is developed in a solid and long-term fashion. The relationship needs confidence and substance behind the annual contracts; both parties need to know that the other will be there in the future. This applies to the customers of the supply chain also. To develop long-term and solid relationships, special attention needs to be paid to these relationships. Since they tend to be dynamic as personnel assignments change, each new person coming into the interface needs to be incorporated into the relationship and to gain trust. Both the operational team and the vendor or customer relationships must be developed in a way that makes the enterprise successful. Not all relationships are required to be strong and personal, but they must be built on respect. The participants must have an understanding of each party's situation, and they must work toward a common purpose. An example of how relationships build and strengthen from a less than confident start is illustrated in the next few paragraphs.

Cooperation between Competitors

In 1984 Congress enacted the Cooperative Research and Production Act. This act, discussed in an earlier chapter, allowed industrial firms to get together and perform research and initial production of something technologically new in a cooperative fashion—even if they were competitors. Initial activity, once the law was signed, centered on the electronic industry. Sematech was formed in the semiconductor industry, and companies cautiously became involved. They worked together to build better process technology to make semiconductor devices and circuit boards. Because this business had been going overseas to non-U.S. competitors for a number of years, the focus was to reverse that trend. After a cautious start, the companies in the industry did commit to having their best people involved and they worked together to improve the industry. The government provided the initial seed money, but industry has kept it going. Cooperating and sharing have occurred with a foundation of common interest and respect for one another among the firms in the industry. As a result, technology and machines were developed that raised the industry to a new level. Another result is a change in direction from devices being produced overseas to growth in U.S. production. The shift was dramatic, and the effect of the cooperation was easily seen. It is a commercial success. The cooperation came from the recognition that global competition was significant and cooperation would provide the competitive atmosphere that is desired in the United States. The cooperation among companies and people at the precompetitive level made the difference.

Additional programs are being catalyzed by initial government funding to try and achieve the same technological change. Both the U.S. textile industry and the U.S. automobile industry are a part of this cooperative effort. In the textile industry it has taken the form of AMTEX while in the automobile industry it is USCAR. These three programs, aimed at making U.S. industry more competitive, required the development of new relationships between people and organizations that historically had been forbidden to interact. It required a significant time to reach a state of trust. Once that was achieved the people were able to work with common purpose toward success. Reaching a state of trust does not happen immediately. The individuals remain loyal to the firm from which they came. However, they must develop a new loyalty to those people who have common problems but had been the competition. The individuals have to become driven by the integrity of the activity and the respect for other participants. Since U.S. commercial activity will have more of these types of cooperation, it needs to be good at understanding the need for building relationships. Relationships do not come naturally, and some help is necessary in achieving trust and understanding.

In Japan, the bringing of industrial parties together for cooperation was the responsibility of the Ministry of International Trade and Industry (MITI), and for many years their activities provided for the common good of Japan in a global competition. They created a capability in Japan that enabled industry to excel. The excellence that resulted made Japan a dominate player in many industrial areas, from cars to cameras. The focus on national competitiveness carried over to other countries in Asia. It also changed how government looked at industry in the United States, and, as described, a movement resulted that allowed for cooperation in a precompetitive fashion as long as it occurred in the marketplace. The consumer was protected by providing continued marketplace competition.

Examples of cooperation have extended even further as enterprises view what they do best and look to others who specialize to perform many of the activities that are not key or core to a firm's success. This type of change in the commercial world involves things like outsourcing or reengineering. It entails changing what has been going on inside an enterprise and selectively getting others outside the firm to provide a service. It is driven by a desire to focus internally on the core competencies. It is assumed, and probably rightly so, that others can do the noncore competencies better than the enterprise. The service provider is probably specialized in the specific area and is much more competent than the firm that is requiring the service. In dealing with the outsourcing or vertical integration in an enterprise, it is important to understand the concept of outsourcing as it impacts mass customization and agility. It is also important to understand how it influences the people and relationships that are affected. When a firm decides to get someone who is more effective to perform a service, it can result in people dislocation. Therefore, it is important that these situations be managed effectively, with the people who have trusted the enterprise being given adequate consideration. The treatment of these people will affect all the others in the firm as they observe the change. If not done properly, it can result in a break in the trust relationship that is broader than with those directly affected. Understanding both the existing relationships and those that are desired after the change must drive the actions with the people involved.

Outsourcing can occur for the critical elements along a supply chain or for those support functions that enable the firm to operate but are not critical to the operation. With the proper concept of manufacturing, various approaches can be successful. Mass customization can exist if the design of the product is right and the capability is set up so that a customized product can be made effectively. It does require attention to the agility of the various capabilities within the supply chain. Building relationships along the supply chain can result in a significant improvement in effectiveness. The concept of "if we want it done right, we need to do it" has been proven wrong. Today, the focus is on the core competencies, and all other activities are looked at as something that others can provide.

The Hewlett-Packard computer-driven printer example given earlier describes a way of designing and manufacturing that is done mostly outside HP. It operates with a philosophy of postponement to make the printer as late in the process as possible. This is an example of designing for customization with a broad product line. To execute the strategy, HP had to successfully challenge the paradigms in the enterprise. It was going to break the rules and the relationships that said HP would make everything it sells. People were cautious with regard to the change, but relationships were developed with the vendors and within the distribution channels to make this happen effectively. Almost every person in the operational supply chains needed to view this direction as being different from the way business was done in the past. They needed to develop a comfort with the design of the product into plugable components. They needed to accept and then promote the establishment of new relationships with the people making the components. They needed to accept that others would be using the HP technology and would have access to it. Further, they needed to develop the capability for the assembling of the printer just before it was delivered to the customer, which involved the changing of many relationships and the development of new ones. A new supply chain and team concept was required. Agility and mass customization arose out of these new relationships. Cooperation along the new supply chain needed to be established and nurtured until effectiveness was attained. To develop as dramatic a change as was achieved by HP, a process was followed to allow the change to occur.

There is a process when one is considering reengineering and outsourcing as a way to make a supply chain more effective. There are rules that provide direction for the activity. The first step is to determine what alternatives the organization has in operating the supply chain or enterprise. To properly analyze this situation, the firm must look at the competencies and determine which of these are core. At this stage, a look into the future is necessary to determine whether a competency exists or can be developed that would improve the effectiveness of the firm. It might even become a new core competency. Both of these self-assessments are important before a company engages in the development of a strategy with outsourcing considerations. Proper assessment is critical in building on the core competencies of the supply chain or enterprise. It is also important to make sure that the competencies that a firm desires are considered. Then it is possible to look for others who are more skilled or effective at these activities as candidates for outsourcing. Next the program of assessing alternatives and building new relationships can start. Outsourcing probably starts with simple things like the cafeteria or canteen. Eventually companies begin looking at why they conduct certain operations critical to the supply chain. In HP's case, it outsourced the building of the modules.

Another example of when outsourcing should be carefully analyzed is in the chemical industry. A firm in this industry, Midwest Chemical, has exclusive and proprietary technology in the chemistry and materials technology of what it makes. In this regard it is a world leader. The steps along the supply chain that deal with this exclusive technology are the firm's core competency. This means that the company would not outsource any of these steps for fear of losing the technology. It is also important that the company does not lose the ability to learn and moves down the learning curve by continuing to improve in the core technologies of the firm. The supply chains within the enterprise represent the place where new products can be introduced. By having the capability within the firm, the introduction is managed and the new product or process technology is protected.

Midwest does outsource *some* of its activity. Outsourcing a purchased raw material that is vital to the success of this firm is done with confidence. The technology advances resulting from research, development, and engineering done by this firm are made available to the suppliers to help make the vital raw material. This provides for an excellent customer supplier relationship with a lot of reasons to trust and work together. The firm also makes sure, by proper selection of companies to outsource with, that capacity is available.

Midwest also outsources things like package materials. These are integrated with the firm's supply chain so the right package with the right graphics can be supplied in a just-in-time fashion. It uses distributors as a prime vehicle for sales. It outsources software and hardware in the Information Technology (IT) area. Other things, from cleaning to cafeterias, are outsourced so the enterprises within the firm can focus on the primary business.

Another firm, A&V Conferencing, which has an integrated audio and visual conferencing product line, contrasts with the preceding examples. It determines its outsourcing relationships by the various products in its offering. The products are integrated electronic equipment that allows for visual and audio conferencing. The key to the technology is the integration of components into a system that functions effectively. The firm outsources many of its simple products in the line. The only requirement is that the supplier put the firm's logo on the product. As the product line gets more complex, the firm integrates components. It buys components from others that are produced to specification provided by the firm. The firm has designed the system and also does the integration and installation of the systems. It relies on its ability to design, assemble, and install the systems. The products with a high degree of complexity are outsourced to a number of component suppliers that are the best at what they do. The firm retains the core capability to design, integrate, and install the components into a product at the customer locations. They also, as in the case of the HP printer, have the ability to design both the components and how they fit or are integrated into a system. This, along with installation, is the firm's core competency.

The latest product of A&V, a sophisticated audiovisual system, has brought the technology to the leading edge and has introduced a new level of complexity. The product has pushed the firm into new markets where the requirements are more severe. It pushes out the envelope of capability of the firm and distinguishes it from the nearest competitor. In this case, the firm chooses to do all of the new technology in-house, developing the various components and integrating them into a working system. In doing this, A&V secures its technology with patents and thus prepares for eventually getting the components made externally. The new core competency is protected by keeping it in-house; the firm will only get components produced externally when the development effort is complete. The new system still requires that components be developed, integrated, and installed at customer locations. The newness of the technology dictates that the company keep the technology development process internal. In this regard, A&V is like the chemical firm example.

Providing high-quality audiovisual teleconferencing products to the marketplace is another core competency for A&V. The product includes making, integrating, selling, and servicing. This firm has thrived on building relationships with its customer and suppliers. It develops these relationships to sell and provide service for the units. It even gets help in making the designs of the products easier to manufacture through a unique relationship. The firm is agile with a limited invested asset base and can respond with new product offerings in a short time, which keeps the technology advancing. A&V is the global leader in its field. Outsourcing is a very important part of the strategy.

The four preceding examples, Boeing, Hewlett-Packard, Midwest Chemical, and A&V Conferencing, each have something in common, although they differ in the way they do business. They each keep their core competency in-house, use others with expertise, and know how to enhance their positions. Each company's core competency is based on exclusive knowledge. The exclusive knowledge allows them to deal from strength in the markets where they do business and gives them the confidence to use outsourcing on those elements of doing business that are not their core competency.

These four companies, and many others, develop relationships with suppliers and customers, and they all share in the success of the products they provide. The relationships are nurtured and considered as an important part of doing business. They are worked on to improve their effectiveness and become very trustworthy. Relationships are vital in agile and mass customization. Each company has developed a capability where focused cooperation and partnerships with others allow them to compete more effectively. This is summarized in Figure 12.1.

Figure 12.1   Key Factors in Relationships

Boeing, Hewlett-Packard, Midwest Chemical, and A&V Conferencing:

1. **Core competencies,** based on exclusive knowledge, are kept in-house.
2. **Other competencies,** where others are experts, are outsourced.
3. **Relationships** based on trust and human interactions are developed with cooperation and partnerships.

The Goal Is a More Effective Enterprise.

This capability includes managing a supply chain across company boundaries with the flow of materials, information, and cash all being effectively coordinated. It is not a simple task, and effort is required to develop this capability. Focused cooperation and relationships go well beyond supply agreements. They need to be based on trust and human interactions at many levels. It is a place where the human organization and knowledge workers are critical to the success.

The Need for Agile Business Relationships

Business relationships are important to optimization of commercial results. These relationships vary and tend to develop as needed, changing according to circumstances. Cultures around the world operate in different ways. Many cultures require building relationships before business can be done. In some cultures, relationships are developed quickly and then discontinued when not needed. In any one culture, levels of relationships exist, from that in the retail purchase, where the transaction occurs quickly, to the long-lasting relationship where trust is developed and many business deals are made because the partnerships are based on understanding and common interests.

Today, doing business in China is the focus of many businesspeople around the world. Through experience they have discovered that the Chinese culture has unique ways and that a lack of understanding can make for frustrating experiences. The Chinese typically look for *developing a relationship* with a potential partner first. Once the relationship is developed, the *logic of the commercial opportunity* can be developed and understood. The last step is *developing the legal relationship* under which business will be done. The strength of the relationship and trust involved becomes the underpinning for the commercial activity. If a businessperson works in another order, frustration will result, and establishing the commercial union will be difficult. The first step, developing the relationship, can take a long time and depends on the type of commercial activity that is anticipated.

The steps in China may not be as unique as first thought. In the Western world we also tend to build relationships while doing business and some of them turn out to be strong and long-lasting. The true partnerships that result in a number of good commercial deals have the strength that is attained in the Chinese relationships. However, the Western approach is more likely to be one of *legal documents*, like secrecy agreements, followed by *logical discussion and concept development*. After this is completed and it looks like something commercial can be attained, the *relationship is developed*. Experience would indicate that the Western world could learn from the Chinese. An effort to develop a relationship before developing a logical commercial deal can build a longer lasting and more profitable interaction. Too often we leap to the deal and don't take the time to understand each other and build trust and respect.

In Europe, the Germans are very direct people who tend to want to develop commercial relationships with precision. However, if one assesses the activities that go into building deals in Germany, one will find they are more like the Chinese than they appear. Time is spent in getting to know each other. It may take the form of multiple visits or even a hunting trip into the forests of Wiesbaden. It is all part of understanding what each party is all about. It could involve sitting in the Englisher Garden in Munich and comparing notes on business and technology, but the real purpose is to chat and understand each other and use that understanding as the basis for doing business. Developing a relationship can result in trust that will make two parties feel very comfortable in working together. If working together is expected to be significant, then it pays to invest in the relationship development.

The Germans like direct discussions and usually say precisely what they are thinking. They thrive on the technical or logic of the expected deal. In contrast, the British tend to want to talk and communicate and take time to get to the deal. It is a different approach but is aimed at the same thing—gaining understanding of each other and using that understanding as a base for the commercial relationship.

In the United States, the approach appears to be to engage in business and if it becomes important enough, then the relationship is built.

Making a quick sale seems to be the dominant feature of commercial activity in the United States. There are examples of excellent business relationships that are based on trust developed over the years. These tend to occur more frequently as firms are narrowing their focus to the core competencies. This results in more outsourcing or purchasing and letting suppliers manage their customers' inventories or provide other noncore activities. Earlier, examples were given of where manufacturing activity is outsourced. How are these relationships built when they are being done in such a quick fashion? These relationships seem to thrive on being agile and getting the deal going. The relationship seems to be based on the knowledge of whom each party is doing business with and what reputation the firms have. In the usual day of the business professional it becomes more of a relationship between the firms than between the people. This enables each party to develop a relationship that is good for the firm and makes sense for the people developing it. There is an important lesson in how these relationships can be quickly developed and just as quickly be discontinued as business changes. It may be something that will enable U.S. commercial activity to become responsive and agile. To do that, the decision to understand how strong a relationship will be needed should be overt. That decision must also include whether a personal relationship between the people involved is needed as a foundation for the commercial relationship. This depends on the local culture and the individuals involved. In the broadest sense it would mean focusing on strong relationships with key customers and suppliers. Various levels can exist for others, depending on their importance to the firm. Just what is needed should be thought out and then cultivated.

The level of the relationship is important in considering just what should be. In recent years, retailers have realized that a friendly and courteous approach is very beneficial to business. They work hard to find the right people and train them in a way they feel gets the best result. It might be an Avis rent-a-car person in Hamburg whom you approach after you realize that your travel agent did not make a reservation. It may be a mail order person at Jackson and Perkins, the leader in providing roses and other gardening plants in the United States, who is taking an order for roses or offering a special to move an overstock. Wherever it is, the positive attitude and friendly demeanor contribute to developing a quick relationship with the appreciative customer.

In other relationships, a company builds an alliance with a supplier that is stronger than a spot purchase, where the supplier offers a special service and caters to your company's needs. The service may be vendor-managed inventory or just-in-time delivery. The supplier operates with the idea that the relationship is important; if the vendor performs to the satisfaction of the customer, then the alliance will be maintained.

Contracts are another level of relationship, and they bring the provider and the customer together with specifics in mind. The contract may include performance expectations and price adjustment formulas. It may include incentives for superior performance that can be earned. Developing the contract relationship requires negotiating until both parties are comfortable with the deal. There is also a time when both parties are getting acquainted and defining what will be brought to the table. The traditional contract was aimed at safeguarding the business. Today's contracts tend to be aimed at making the business grow.

Another relationship that usually takes more time to develop is based on each party bringing something to share with the others. A partnership is a more serious development for those involved *because* of the sharing of capability. A partnership is a relationship that is expected to last a long time and is thus built on interdependent reliances. Both firms must have an important need to partner, and the relationship involves developing a position that is based on this need with trust. A contractual relationship probably results, but there is more in this relationship than just buying and selling. Technology and joint sharing of what is developed may be involved. The relationship may involve more than two parties as participants. It is intended to make something better because the parties are working together.

Some relationships develop over time where an agreement exists that if a certain activity is to occur, then it would be done with both parties. Others might be capable of becoming the partner but they are excluded because of the trust built between the two parties. This type of relationship might include the understanding that a vendor will follow the customer to all parts of the world to supply raw materials. The vendor might be informally committed to building an interdependent manufacturing facility at the customer's fence line. The agreement could be to package components or critical raw materials. It could be the exchange of a feed stock from the customer to the vendor with return of the raw material and a by-product for reprocessing. The relationship can range from the simple to the complex. Because of the trust between the parties the agreement need not be in writing, and only when the deal approaches do specifics need to be defined on paper. It is a strong relationship and each party treats the other in as fair and respectful a fashion as possible. Negotiations can still be arduous, but both parties trust that a deal will be struck.

The strength of human interaction is a vital consideration in developing the relationships that one desires for doing business. The speed with which they need to be developed is increasing. More awareness is necessary to determine how strong of a relationship is required to make the commercial deal work. Its definition and execution need to be the result of decisions about what is desired.

The world of agile response and mass customization will require developing relationships that support the response to unexpected change. The parties in the relationship must have or must develop the capability to respond rapidly to the needs of the customers and do it in concert with each other. Both parties must be aimed at the agile response and/or mass customization. The next part of this chapter on agile relationships will deal with the development of an organization that can respond. It will address the important role of people in the organization and making sure that the atmosphere exists that allows them to make the agile or custom manufacturing firm a success.

Agile Relationships within an Enterprise

An enterprise balances the needs of its customers, stockholders, employees, and society at large. This is done in a fashion where the needs are integrated, compatible, and acceptable to each of the four groups. This balancing establishes the base for developing the business strategies that a firm uses for its supply chains. At this stage, another act of balancing is required. The firm must balance its capital, finance, technology, and human organization. In previous chapters we have discussed the first three and only touched on the human organization. In the following paragraphs the discussion will focus on the people in the organization—it is the people who make the organization work.

An agile enterprise must deal with a number of critical elements to develop the capability to execute its business strategies. The first of these is the *workflow processes*. This is the definition of *how* an organization will conduct its business. It is not a single process but a group of processes that, when integrated, make up the activities of the firm. Workflow processes must exist for every activity, from how a new product is commercialized to how people are paid. They are put together to make activities more understandable and reproducible. Each process is intended to consider the people who are conducting an individual process and ensuring that the tasks within the process are as efficient as possible.

In optimizing workflow processes, attention is given to whether each step in the process is necessary for the final result and is done as efficiently as possible. The people in a workflow process are asked to improve the effectiveness of the process. They must not only understand their tasks but also their role as part of the process. That role is usually one of not only doing the workflow but also improving how it is done. Information systems are critical in operating the firm, and the workflow processes must be integrated with the information system. They must be balanced with the flow of information. If the information system requires a lot of handoffs, then the people must be structured and trained to do that effectively. If the system works with information entered once and used by all, then the people required to process the information are probably minimized. The workflow processes must be designed for the situation, and the people must work to improve it while using it.

An agile organization also needs to focus on the capabilities people need to run the firm or a supply chain. Having the right *skill* in place will make the firm more productive, and individuals will make a bigger impact. People with the appropriate skills can be hired into the organization, or those without the skills can be trained before or during the work activity. Different activities, and indeed different supply chains, require an understanding of which people capabilities will be the most effective. For example, a commission selling activity requires a different type of person than one who works to build relationships with the customer. Some organizations work best under time pressure. People who thrive in this type of environment should be matched with the jobs. High-technology or scientific organizations require people with those skills. Matching the right people for the right jobs and then providing the right training are critical in a human-oriented organization. Indeed, they are key attributes of this type of enterprise.

In an agile enterprise, attention must be given to *who* will do what with attention to the workflow processes and the skills required to be most successful. Establishing an organization that is very flat, with decisions and direction occurring as low in the structure as possible, requires a definition of who is responsible for which task. This must be done with more definition than in an organization that has more layers and is hierarchical. In the hierarchical organization the work is defined so that a protective interaction occurs between people to ensure that the right thing is accomplished. This exists less with the flatter organization. The workflow processes and skill requirements may be the same, but just who does what will be very different.

Another element to a human organization is the *mind-set*, or *culture*, that determines the environment in which the enterprise or supply chain operates. Many elements make up this environment, and they can vary from rewarding people for risk taking to people having a fear of trying anything new. Either one might be appropriate, depending on the nature of the firm or department in the firm. The leadership in the firm plays a key role in establishing the culture. The organization develops a fit to the management style and the actions of the top people. Other elements of the environment might be the job security associated with different firms and their culture. Some firms use the "shadow of the gallows" to motivate the workforce whereas others work from positive reinforcement. The concept of the shadow of the gallows can be either external in nature or internal. When a firm is facing a competitive threat, it is external. When it is used as part of the culture, it becomes internal and a way to do business. Many things make the culture of a firm what it is. It is an element of the human organization that can be managed and thus is subject to decision making at both the strategic and operational or tactical levels.

The human organization must understand and decide how things will be done, what skills are required to be most effective, who will do what, and what the environment will be to provide the direction for the enterprise. These elements, when combined with the capital, finance, and technology, compose the strategy of the business. The direction provided by the preceding elements takes the form of tactics or initiatives.

Some of the initiatives come from the policies that are developed for the enterprise or supply chain and include things like employment policy, job staffing levels or requirements, compensation policy, benefits, and career or skill development. There are also initiatives around the way the corporation will conduct human interaction. Management behavior has a lot to do with how effective a corporation can be. The managers must "walk the talk" and demonstrate the type of actions that are desired and what culture or mind-set is desired.

The consideration of people as part of an agile enterprise is one of the most important features of attaining success. Many top managers spend most of their time on this type of activity. Making sure that the right people are in the right job and that the culture is what is desired becomes almost a full-time job. The right people and mind-set, with good business strategies, will make impossible goals attainable.

People Make the Difference

Agile and custom manufacturing supply chains require that people become the critical part of responsiveness to become effective. They must be receptive to operating in a way that initially may not be comfortable. They need to develop a confidence that what is required can be attained and that responding agilely or in a custom fashion can be made easy. The human organization must ensure that people have the right skills and motivation. They must be given the ability to be effective and fit into the direction the firm is taking. The culture must support them so that they can perform. The culture of the company must also reward the performance that is desired. The performance that makes the organization successful must also be the same performance that makes the people successful.

Historically a firm was capitalized and the assets became the tools that people used to produce. Henry Ford provided the place to work and the devices or tools that were needed to produce the automobile. The person or worker was required to toil and carry out the tasks. The concepts for how the work would be done and what materials and tools to use came from the engineers, who were the knowledge workers of the time and understood how things should be done. As the industrial age has progressed, more people have become knowledge workers and have the expertise to make a firm what it is. Today's knowledge worker has developed more mobility and can take expertise from one firm to another and apply the skill to another firm's problems or opportunities. This trend points out the importance of people in the enterprise at all levels. The operator may be the only one who knows how to effectively produce the product. The engineer knows how the parts fit together. The salesperson knows the customers and what it takes to make them happy and purchase the product. This knowledge composes the true asset of a firm and, combined with the capital-related assets, make a firm successful. The recognition of this has changed how people are treated; things like compensation programs have changed to ensure that the knowledge worker remains with the firm.

Changing to a different way of doing business will be resisted. It is a natural response. Each individual will have his or her own reasons and logic about why change should be resisted. A few people find change enjoyable, and they usually become the change agents who thrive on making something different. Others will look at change as anything from a personal threat to a frightful unknown. They may have past experiences where organizational change has left a bad taste. Work overload, lack of change resources, or not being included or involved will cause resistance. To make change successful, the vision of the change needs to be understood and the incentives clearly defined. In many cases, understanding why change is needed and what the change entails is enough to make people a part of the change team. In other cases, information may not overcome the threat that the change implies. Different actions are then necessary to bring the reluctant people onboard. All concerns of the people affected and involved must be addressed. They cannot be overlooked because the firm's people are who will make the change work and the benefits occur.

The people in an organization will not only need to be motivated to change. They must also be committed to making the change work and providing the fine-tuning of the change as it is put in place and new or unanticipated situations arise. They must find the solution that keeps the improvement moving and, hopefully, makes it even better than originally conceived. This typically occurs while everyday duties are being performed. Human beings are uniquely able to cope in an agile way to the unanticipated events that occur as a significant change is rolled out, which makes them invaluable in the world of agility. People provide the intellect to consider all the necessary details and make the general direction of change a success. They make it happen, and thus an agile company must ensure people are able to perform at their maximum. They must be treated fairly and their importance recognized. The credit for success must be earned and recognized. Giving credit to the wrong person for an accomplishment will result in a breakdown in positive relationships—fairness must prevail. Nothing is more detrimental to maintaining a performing and motivated workforce than unfairness. Fair play is something that the average worker, in most cultures of the world, is schooled in. When something is not fair, people will lose motivation and move toward resentment or apathy. This will result in the loss of all those little things that make activities effective.

Thus, care must be given to how people are treated. Steady employment at an acceptable wage is the minimum. Once that exists, effectiveness can be enhanced by providing an atmosphere that positively rewards people for their contribution. A company must also make sure that good deeds are always recognized and exceptional performance commended by both peers and management. The focus must be on the dignity of the individual, and recognition is a key ingredient.

People must also be in the right job so that their talent fits what is being expected of them. Training and skill development can stretch the employee to fit a job that requires more skill than initially available. Expectations regarding both the learning process and the job must be understood by both the team leader and the employee. Progress on acquiring a skill must be measured and frequently discussed. As skills develop, feedback is essential to maintain progress. The goal of coaching is to make employees be and feel successful. Successful organizations are made up of successful employees. They go hand in hand. It is also important that the consequence of not being able to develop a skill is understood and that a positive alternative is available. Managers and leaders must envision the needs of the organization and then ensure that needed skills are developed or brought into the workplace.

An example of the importance of people is in the chemical industry. There is a very slight probability that chemicals could be released during processing. The assessment of the situation has resulted in the industry's improvement of the physical equipment to the extent that it is relatively fail-safe. In some cases domed enclosures surround an entire process. These domes are people-free when the process has chemicals in it. They are used to contain the last trace of any material that might escape. The operating experts have concluded that human error has become the most probable cause of a release; the probability is that if a release occurred (which is very low probability), it has a 90 percent chance of being due to a human error. This is not a reflection on the individuals involved but only a place to focus attention to mitigate the chance of failure even further. This shifts the focus from the equipment to the way the people operate it.

The people-oriented techniques that have been developed to address this situation have been focused on ensuring that the procedures are always followed and are right for the situation. One firm in the chemical industry in Europe assigned an operator (in a rotating fashion) to observe fellow operators. When a procedure is missed or not followed, a correction is made and learning results. Both the operator making the correction and the operators running the chemical plant learn the procedure and determine whether the official one is the best.

This becomes a built-in improvement process. It is somewhat akin to what happens in the pharmaceutical industry where a person checks the steps of a drug's production to ensure procedure is being followed. Because of the very low probability that any incident will occur, improvement is measured by compliance to the procedures. It stresses the importance of people in a fully automated chemical plant. People are a key factor in the safe operation.

The people in the organization are a valuable asset that must be invested in. It is not enough to just hire people and hope they can perform to the expectation that an enterprise has established. This is unfair to the individual and to the organization. The culture must stress that learning, skill development, and training are a key, ongoing part of becoming and remaining a productive employee. People will appreciate it, and they will become dedicated, motivated, and effective employees. The process does not end with the initial hiring. The world, the company, or the supply chain in which business is done is a dynamic environment, and change must occur. The people must be viewed as needing to develop new skills throughout their career. The opportunity must exist for the growth. It's part of the agile environment that is needed in the enterprises of today. People are the key part of that enterprise.

The next chapter will develop the pathways to reaching an agile and customization enterprise. It will spell out what needs to be done to achieve a result where an enterprise can respond to unexpected change in a rapid fashion.

Summary of Relationships

This chapter has highlighted the need for relationships as a cornerstone to an enterprise's success. To be successful there need to be trust and loyalty between customers, suppliers, and employees. These relationships are a critical part of doing business. With the advent of mass customization and an agile firm, the relationships will need to be developed quicker and last a shorter time. Customers will get a custom-produced product and then, with the need satisfied, disappear. The product line will change quickly and thus the relationship with suppliers will also need to change. The culture and organization in the enterprise will need to be dynamic and agile as the human side of the enterprise works to be competitive.

CHAPTER 13
Pathways to Agility

To make the significant change from the way we do business today to a way that includes agility and mass customization will require a change process that follows a pathway to agility. My experience is that progress will not be made toward improved operations without the use of a disciplined and organized process of managing the change. The process of change in a supply chain or in an enterprise incorporates a significant number of best business practices. At first, these practices will not be well understood by the organization and their benefit to the organization will not be easy to articulate. Leadership must take the time to understand and learn improved business practices. Then a change process or pathway is essential to provide the method to make the needed improvement a reality. These conclusions come from various efforts where I worked to make changes. They did not succeed until the leadership saw the importance of the change and unless a process was followed that managed the change along pathways. However, after many tries the change process has become manageable.

Change in the commercial environment and the way an enterprise does business does not occur easily. It requires a significant effort. It is something that will occur in a continuous fashion and thus needs a well-defined vision or direction. Defining the vision or direction of the change is a formidable task that requires forethought and a well-developed approach. Determination of where improvement is needed and what the improvement might be is the responsibility of the corporation's leadership. They provide the framework for change by establishing the concepts for what business the firm will be in and what culture will exist for carrying out that business. They will also determine the level of benefit the firm will expect to be successful. Since the desired economic returns dictate many things in the change process, they must be stated in language that can be understood and measured.

The economic benefits determine the rate at which the firm needs to grow and what the level of profitability must be. This overall direction must be consistent with what is feasible. It does provide the foundation for the change program. The leadership of the firm must approve the activities consistent with the overall vision and direction. They also must provide the change process and the discipline and motivation needed to achieve the desired results.

Establishing the Direction of Change

The leadership of a firm or supply chain is usually the owners and the top few officers who are running the operations. The effort needed to define what the owners and key leaders would like to achieve with the firm or supply chain is an important effort. It requires forethought and structure in developing the appropriate directions. It is not enough to conclude that one keeps doing what has been occurring. If the status quo is desired and leadership does not provide a direction, the change will exist in the corporation at various locations without overall harmony or coordination. Priority for change will be determined at each of the many centers of change on an individual basis. Thus, change will occur, but it will not be effective; it will approach random change. It will not be focused, coordinated, or integrated. Therefore, a vision and direction must be provided.

It is also important to have agreed-upon processes or pathways for change. By having the proper pathways, change can occur in unison throughout the firm. The process must include times of implementing change and times for reassessing direction. This needs to be done at each level within the enterprise. Periodically, possibly every three to five years, the overall charter and direction must be revisited by the leadership. This is required to assess the improvements and make sure the right course is being followed. Reassessment should also be a time when the vision and direction are confirmed or modified.

In the overall process of pathways to change, from direction of the leadership to change actions, various elements must be included in the process. These include shifting to streamlined supply chains or enhancing the introduction of new capability. The change that affects the way the company runs is different from change that is associated with what the company does or what business it is in. One is driven by the operational environment with emphasis and focus on doing today's business better. The other focuses on what the company is offering in the way of new products, processes, and markets. The two elements combine to form the whole change program for the enterprise; they will be divided into a pathway for operational change and a pathway for new product, process, and market capability. The two pathways operate in parallel, with a harmony consistent with the overall direction of the firm. Agility and mass customization will be discussed as a part of both pathways or programs.

Pathway to Operational Change

Operational change is a continuous process that is driven by the need to improve. The change is accomplished through either continuous improvement or breakthroughs. Each can be managed but in slightly different ways. Movement to world class in a continuous improvement fashion requires that all the people involved in the improvement process are aimed in the same direction. This means they must be focused on a vision of where the supply chain or enterprise would like to be. The technology or best practices that are the subject of change are usually easily available. They come from various people who have seen best practices working at other companies and feel that they would be useful and beneficial in the way the firm does business. The type of change that will work best initially is continuous improvement. It does not require breakthrough thinking but the application of things that are demonstrated by others to be effective.

In contrast with continuous improvement is change that incorporates breakthrough thinking. Breakthrough thinking may even require technology that explores the unknown. In most cases of breakthrough change, the thinking is not well understood and specific actions do not flow easily. It is driven by a vision that expands the envelope. It pushes to new frontiers, perhaps even pioneering and going where no one has been before. Both types of change are essential to the successful transition from a firm's world-class current state to an agile and mass customization future state.

The process of getting from where we are to where we want or need to be does exist. It is multistep with recycles. It starts with an understanding of where the firm or supply chain is and determines whether this status will be satisfactory in the future. In most cases the status will not be satisfactory and change must occur. The gap between today and the future determines the speed at which one must travel. This determines whether directed continuous improvement is the right approach or a breakthrough change is warranted. The gap needs to be assessed with a future competitive situation as the driver. Firms that have made significant change were driven to a future by the leadership that was outside the range of thinking of many people in the firm. Gap analysis does not always need to provide a major change in what a firm practices. The major change is probably not used often enough to properly enhance the competitiveness of the firm. The following example describes a process when a directed continuous improvement is required.

The Specialty Resin and Lubricants (SRL) is a supply chain that does $140 million dollars of business. It has a physical volume of 26 million pounds of product a year that sells for $4.50 to $6.50 per pound. The gross margin is 78 percent. It has an inventory in the whole supply chain of $4.8 million dollars that represents 90 days. Customer service is at

95 percent to customer request date. Growth is low, and the product line has existed for more than 20 years. The following process was used to make the supply chain a much better contributor to the corporation.

The leader of the supply chain brought together key people from the supply chain team. Their initial role was to define the supply chain and who the people on the team were. They also defined who was not a part of the supply chain and tried to make as many of the jobs as possible ones that dedicated the people to the supply chain. Interfaces with other supply chains were defined, and it was determined that about $280 million dollars of other business was leveraged by the intermediates and products from this supply chain. This leverage was important for the success of the firm and thus became a very important part of the change process that was anticipated.

With the supply chain and the team defined, a process was initiated based on making improvements and working together. This process was put in place and through super cooperation the performance of the supply chain was dramatically improved. The investment in inventory along the supply chain was reduced from 120 days to 90 days with the resulting savings in working capital. Capacities of equipment were defined, and better utilization enabled customer service to be improved significantly. Performance to the customer request date went from 74 percent to 95 percent. The supply chain team integrated itself along the 10 steps involved in manufacturing the product line. Every team member became an advocate for the customer.

After the team building, and the success that was initially attained, it was felt that a reassessment of direction was needed. The leader of the supply chain brought together a few of the team members, the two business managers who were responsible for the long-term health of the business and the operational leadership that provided a home for the supply chain. They started the effort by reviewing the vision and direction of the corporation. They also reviewed the statements of the culture that was expected for all supply chains. The intention was to make sure that the eventual actions of the team were consistent with the mores of the corporation.

The next step was to develop a mission statement. The mission statement applied to the fundamental unit of business management—the SRL supply chain. It existed inside the mission statement of the corporation. In the SRL mission development more of the team from the supply chain was brought together. They reviewed the findings of the leadership team. This was followed by the leadership and the supply chain team working together to improve the definition of mission. The product that resulted was agreed to by all. It was a collaborative document and formed the basis for the business and team direction. The time horizon for the mission was three to five years with annual reassessments. This time line was viewed as necessary so that change could be prioritized and an implementation schedule devised consistent with the capability of the supply chain team to manage and implement change. It also enabled technology that had a longer horizon for development to get an early start and be available when it was needed in the improving supply chain.

The mission statement read as follows:

The Specialty Resin and Lubricant Supply Chain Team is dedicated to optimizing the flow of goods and services between suppliers, steps of the supply chain, and our internal and external customers while optimizing profitability and utilizing mature product technology in mature markets in ways that provide life cycle extension.
- Emphasis on extending the product life cycle of a mature product line
- Emphasis on extending product offering and delivery form
- Emphasis on process viability and maximum use of assets
- Emphasis on regulatory trends that necessitate investment
   The Business Posture was deemed—Mature

The development of the preceding mission defined the nature of change as continuous improvement. Implicit in the statement was a desire to significantly improve the performance of the product line through application of technology and techniques. This was addressed when the team developed the critical success factors that dealt in more detail with things like product service and quality. Ideas were developed of what actions needed to occur and what the resulting improvement would be in a series of measurements. Other critical success factors were cost reduction, supply chain responsiveness and reliability, environmental technology, and safety. Each of these areas, including the improvement of the performance of the product line, was developed with suggested specific actions defined. At this point, the preliminary change program was put together and the team looked into the corporation for further help in getting the change schedule prioritized and the specific actions implemented.

To do this they involved the areas of expertise within the corporation. The team was looking for the best practice, and the leaders of the areas of expertise are continually benchmarking other firms' performance. The expertise was in areas like purchasing, logistics, packaging, utilities, process mixing, process separations, process control and information, environmental technology, and analytical testing. Additionally, the experts who provide the process and product technology for this product line were involved to ensure their plans were consistent with the supply chain team's plan.

After a couple of months of discussion and prioritization, the five-year plan was determined with specific emphasis on the current year. In the plan the role of the team members and their areas of expertise were defined. All parties knew who would do what and when it would happen. This is the stage where agility and mass customization came into play. Both were concepts that were included in the plan, and actions were outlined that developed this capability for this supply chain.

The purchasing organization pushed for a vendor-managed raw material inventory with consignment. Raw materials would be subject to a make versus buy assessment as a routine part of the procurement process. Other things that were included were cycle time reduction and a shift toward make-to-order from make-to-inventory. The scheduling capability would be integrated across the supply chain and done by a few schedulers located together. Besides scheduling the supply chain for delivery of material, they would schedule the preventative maintenance activity to control when equipment would be out of service. A program was reinforced that pushed the products and processes to be more reliable by gaining more knowledge. The knowledge was then used to improve the process and ability to produce the product consistently. Specifications were improved with a number of tests reduced significantly. Variation of product was a key focus for reduction. Parametric acceptance would be worked toward but could not be accomplished during the first year.

The information system for both business and manufacturing was integrated and improved. It was tailored to the responsiveness that the product line would require when business was operating in the make-to-order fashion. The manufacturing processes in the supply chain were assessed, and those nearest the customer were the focus of developing a custom and agile customer response. Since customer daily demand would provide the demand signal, it was necessary to analyze this and make sure enough capacity existed in either the units dedicated to the products or new units. Many situations required developing a primary and a secondary source for where the product could be made. This agility would allow customers to be satisfied even when surges in demand for a product occurred. The product offering was to be tailored so that combinations of properties could be specified and the system could produce a customized product for each customer. Custom coloring of lubricants or resins was also being explored to further increase the offering to the customer.

The first year of improvement was a continuation of the initial efforts of the team. Significant and measurable changes occurred. Most of these were internal programs that helped to reduce the supply chain lead time from 90 days to 60 days, with a target of 30. The external changes that dealt with the supplier or vendor were all successful, with vendor-managed consignment inventory becoming a reality. The vendors kept the tankage inside the plant at the desired level, and payment was made when the material was used.

The customer improvement program had mixed reviews. Those parts that delivered less variable product quickly were greatly appreciated. New offerings of products that were produced to tailored results were not as well received. The process of customers changing the way they did business seemed to occur more slowly.

At the end of the first year of change, the reassessment process was an exciting, high-energy interaction. The supply chain team wanted to accelerate the change process. This put unexpected demands on the expertise centers, and some adjustment in thinking was needed. The expertise centers had always been the pushers of new things, but they were now in a different position. The whole team felt that it was on a very beneficial improvement cycle that would result in a much more competitive supply chain. Customers provided just enough positive feedback to indicate that the agile and mass customization part of the program will experience success.

Agile response was seen in dealing with the suppliers or vendors. They showed a real desire to respond. Internally, agility was attained in the scheduling activity and the qualification of primary and secondary production sources. Mass customization, which was too new for the customers, will need to wait to be put in practice.

This actual change activity reflects the steps necessary for making a supply chain more agile. The change occurred, and the reassessment process has been proven very useful. The pathways or change process worked. Commitment was received, and everyone knew what was expected of them and what others were going to deliver. Success occurred in 80 percent of what was chosen to be accomplished the first year. Why things were not accomplished was analyzed. This knowledge, combined with the successes and the five-year plan, was used in the reassessment. The change process or pathways will be summarized in the next few paragraphs.

The initial step to creating an agile supply chain or enterprise is to identify that an increased *customer focus and responsiveness* is desired. This can be done either for a supply chain or for the whole firm. It is an important step in that one must understand the existing customer relationships and determine that new ones will be beneficial to both the customer and to the firm. It is a decision that the leadership of the firm must make in its overall effort to improve effectiveness and competitiveness.

Once the need is established to change, then the *supply chain and its operating team* must be identified. This part of the process requires commiting of people to the chain while having others reassigned when the supply chain is only a part-time job. A leader must also be established who has current business and operational responsibility from vendors to customers. The team and its leader must have full responsibility for what happens in the business-oriented supply chain. The customers must belong to the chain. The vendors must be team members as a step along the supply chain.

The team needs to initiate change or improvement activity immediately after being identified. This starts the process and the "low hanging fruit" can yield some positive quick results. The initial *improvement activity* needs to occur to get as much benefit as possible from the now integrated supply chain. Significant improvement should be possible in a short time in customer service, supply chain lead time, level of inventory, and supplier relationships.

With the team built and success under its belt, a more formal foundation can be built for the supply chain improvement process. The leadership of the supply chain needs to establish with the leadership of the firm *an understanding of the vision and culture* that will dictate how the improvement process should operate. From this understanding the supply chain leadership will ascertain what performance is expected for this entity. It will also understand the drivers and methods of change; what is acceptable and desired practice is very important to understand.

The next step is to combine the business leadership group with the supply chain team and develop an action-oriented *mission statement*. This statement should not be modified to sound good but should retain all the action-oriented words and the constraining comments. It can be a number of pages long and contain key areas of change and critical success factors. It must be the basis to drive the change activities.

The supply chain team should then put forth a program of *initial change programs*. This is done to provide an atmosphere for change and to define what type of change is desirable. It also establishes that the operational supply chain team is receptive. The change program should be viewed as a five-year program with annual reassessment.

With the initial change program and the mission in hand, the supply chain leaderships works with the expertise centers in the organization to determine feasibility of the program and what other projects should be considered. The intention would be to brainstorm all improvement ideas and test them against the mission and initial program for validity. This is an interactive process. Each expertise center needs to promote the merits of its desired changes. The supply chain leadership must encourage the input. This phase of the activity *perfects the change program* and includes input from all participants with all ideas being considered.

Once the input has been captured, the supply chain team, in concert with the business and corporate leadership, develops a five-year *change program with the first year defined in detail*. This requires the determination of which projects are desirable and what priority they should have. The team then communicates the program to the expertise centers and the remainder of the supply chain with responsibilities and time lines established.

The team and support people then work to *implement the first-year program*. With the responsibilities and commitments in place the process should go smoothly, but as difficulties arise, the direction and mission need to be revisited to confirm that the course is the right one. Obstacles that are encountered must be overcome. Results must be attained.

At the end of the year, the *results of the first year effort are assessed*. They are compared with where the team thought it could be. It is very important at this stage to understand the root causes of both the successes and the failures. This understanding forms the basis for the next step in the process.

With the results in hand, *the five-year program is reassessed*. The mission and directions are evaluated and whether they are still the desired course of action is determined. Modifications are made, and the new five-year program is built, with emphasis on the new first year.

The modified five-year program and new first year are *reviewed with the leadership* and then implemented. The second year of change is then initiated and the implementation of change occurs. This process is recycled every year. The success of each year will allow for bolder activity to be considered. At the end of the third year, when the *effectiveness of the supply chain has truly been enhanced* and competitiveness improved, then a different direction is possible.

The leadership looks at the improved supply chain and develops initial thinking about what the *ideal future could be*. This starts as a brainstorming exercise, but its intention is to build a new direction. It is not as rapid a process as before because it will require more understanding of when some of the ideas will be breakthrough thinking. It will also challenge the direction that was initially established as a foundation for the mission. This part of the process is a very important step. It is the transition from a continuously improved supply chain to a more dramatic change. It challenges the corporate paradigms and presses the envelope to new horizons. With the improved supply chain forming the basis or the platform, it is time to consider the future with breakthrough ideas.

This is the *entrepreneurial brainstorming* phase of the change process, the part that allows the exploration of alternatives that can result in the ideal future. This phase of the activity takes time and all ideas need consideration. Once the ideas are sorted through, then the leadership challenges the supply change team with the ideal future. This is also a brainstorming activity with ideas encouraged. Everything is captured. When the ideal future is brought into consensus, it becomes the driver for the new mission and directions. Change actions are developed into plans, and implementation toward the ideal future begins. This step can be the breakthrough change that includes heavy dosages of agility and mass customization. The stage was set during the earlier years, and now more things are possible. The supply chain team has learned to thrive on change and deal with the obstacles that try to limit it. Eventually the capability will be built so the supply chain can thrive on unexpected but anticipated change. This is possible because the supply chain team built the capability and it now exists.

The process pathways for the development of an organization that can manage change and include it as a capability are ready for making significant change in competitiveness. The second phase of the change process is to make the change more aggressive. This is an important element in the pathways to agility. If an organization does not thrive on change, bold and creative directions will be difficult to initiate. *Prepare for breakthrough change by learning to thrive on continuous improvement. Anticipate and drive the development of the capability to thrive on unexpected change.*

The focus will now shift to the new technology part of the change program. This part of a disciplined change program will deal with new process, product, and market technology and the role that agility and mass customization play in its successful implementation. It will look at change in product line, process, and market as a parallel change process. It complements the operational changes or the continuous improvement activity. New process, product, and market technology was required during the operational improvement process. These new technologies are even more important in the more aggressive change in the breakthrough phase of increasing a firm's or a supply chain's competitiveness. The next part of this chapter will highlight where change needs to be made in a bold and creative fashion and the risks associated with this type of pioneering.

Pathway to New Technology Change— Product, Process, and Market

Most changes associated with new technology in the product, process, and market are manageable using the pathways just defined. They come from the assessment of the situation and are determined during the mission development part of the pathway. The definition of the role of changes in product, process, and market is a part of the improvement process that the supply chain teams develop. These changes tend to be the short-term goals of the new product, process, or market. The more bold and creative changes come in the second phase of the change processes. Although they may be only in people's minds when the initial activity of change is undertaken, the atmosphere must be right for the bolder steps to become a part of the program. In many cases this part of a change effort is where progress toward an agile or mass customization activity can be initiated that could eventually change how the supply chain does business.

When dealing with the initial change to the product, process, or market, it is easiest, and least risky, to work with only one of the three dimensions. Introduction of new products that set the stage for agile or mass customization, which are produced on the existing process and sold into an existing and understood market, will simplify the change and make it easier to accomplish.

If a new manufacturing process were needed to achieve success of the new product introduction, this would further complicate the process of change. The complexity may overwhelm the initial change activity of the supply chain team. The strategy might be to develop the product and process while the supply chain team completes the initial change or improvement effort and then introduce the technology in a structured fashion into the second phase of change for the supply chain. The management of how much change a team that is busy with today's business can facilitate is an important feature of the change process. Too much and nothing will happen. Too little and progress will be slow, and the patience and persistence of the team will be stretched thin.

To be successful in introducing a new process, product, or market, a company must view it as a process of generating knowledge. The needed knowledge is not perceived as the same for all parties involved with this type of change. The new product people feel that a few pilot runs establish the product as manufacturable. The manufacturing people would like enough product to be made to bring the product and process under statistical control with most of the root causes of failure eliminated. These represent two levels of knowledge along the path to a successful commercialization. This concept is depicted in the graphic titled *Knowledge Points-Concepts*, which comes from the book *Going Virtual* by Grenier and Metes.

In Figure 13.1 the knowledge points indicate the following:

A = What you (the team) think you need to know to complete the change or project
B = What you really need to know
C = What you think you already know
D = What you really already know

The correlation of competency with time in a change process defines the development of knowledge and understanding that must occur when making any change. It highlights a misestimation of what is needed to be known and what is currently known. In each case experience would say that we underestimate. We think we know more than we do, and we don't think we will need to learn as much as we will have to. Success requires that the development of knowledge become the prime enabler for change.

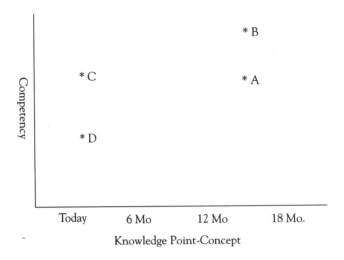

Figure 13.1 Knowledge Point-Concept

In the case of a new product, this situation does exist, and the participating team members must agree on how success will be defined and who will be responsible for generating the knowledge. In too many cases this understanding is not developed, and the team members are stressed when failures appear with no one around to search out the root causes. If the team is operating, then the problems that appear will become everyone's problems, and the solutions will be developed. In new product design this constitutes the concept of design for manufacturability and, in most cases, is a critical element to success.

The same level of knowledge development is needed for a new process or a new market. In the case of a new market using an existing product, a significant amount of knowledge about the market and the application does not exist and needs development. In some cases, a partnership can be formed with the targeted customers and the application developed jointly. In other cases, the producer must learn enough to intelligently introduce the product. Sensitivity must exist for things like package graphics, channels to reach the customer, means of payment, ultimate fate of the product, the requirements of the application, promotional approach, location of the buying influences, and so on. These are not simple things, and the tendency will be to underestimate the requirements of doing business in the new market. To gain the knowledge, firms use test markets to introduce the product and measure success before going further. This helps to reduce the risk of the commercialization.

New processes require the development of knowledge before they can be introduced successfully. The process is not much different from the new product or new market. Knowledge is gained about the capability in a progressive fashion through steps. A frequent approach is to build a small-scale process unit and do test product runs. With the learning from this step, the process is scaled upward to a pilot plant or a prototype line where additional features are added and more knowledge is gained. Manufacturability of products on the new process is continually assessed. Benefits for the new process must be determined and understood. It may be new capacity with or without new capability. The process must show the benefits needed for the commitment to implement.

The steps in making the change where a new product, a new process, or a new market is introduced are similar to the change process for the operational activities and the supply chain team. The magnitude of change is such that it can be incorporated into the existing activities using the overall direction from the team leader. Change should be viewed as one of the activities that will enable the mission to be achieved. It should contain technology that will allow for the supply chain to become more agile with mass customization as a focus.

The world gets more complex when all three—a new product, process, and market—are involved in a change. Then an effective implementation requires a significantly different effort. There is no existing supply chain team. There is no existing mission aimed at improving all parts of the supply chain. There is far less knowledge than required to successfully approach this bold and creative opportunity. The rewards for success must be high because the development effort will be large. Even with this understanding, many of these types of changes occur every year in industry. There is a suggested process for implementation of this type of complex change.

The *new product, process, and market change process* starts with either an internal technology looking for an application or a perceived need looking for technology to solve it. Both work and require a discipline in gaining the knowledge to adequately determine whether to explore further. This cannot be done casually since the error is being overly optimistic. The internal technology champion is driven to see the brainchild commercialized and thus is wearing rose-colored glasses. The perceived need is not well understood and the need to gain more knowledge is underestimated. Both of these need critical scrutiny. The decision will put the firm into action with a commitment of significant resources. This *market assessment* step builds the foundation for the eventual commercialization.

After enough knowledge is developed on the application, the firm or business team must look to see what technology will be used in the potential new product. A product concept must be developed with candidate technology proposed. The proposed technology will determine the functionality of the product. The team must continue to assess the market to determine if the product meets the needs. If the technology does not supply the functionality that is required, then a further technology search is warranted. If the technology exceeds the need, then the organization must determine whether the application will bear the cost of the superior functionality. The process gradually narrows the technology. As the technology narrows, it is very important that the state of the technology is assessed because people will overstate the degree of development of the technology. Careful consideration needs to be paid to where the technology is today and where it needs to be. Too big a gap can result in product failure. The process of applying technology to the potential product and then determining the state of the technology will result in a conceptual product definition. The *product technology determination* is a vital step in the development of the new-product, new-process, into-a-new-market-type of commercialization.

The next step is to develop the new product and the new process in parallel with frequent interaction to determine manufacturability. The market knowledge must also be further developed to test the concept of the product. To accomplish this the business leadership must put an interdisciplinary team together with the task of developing product, process, and market knowledge. It will involve looking at alternatives of product, process, and market concepts. How well this step is executed will determine the success of the commercialization. The interdisciplinary team needs expertise from product development, manufacturing, and marketing. Each person on the team should have allegiance to the commercialization team and the business leadership. They are responsible for taking the concepts and turning them into real products, processes, and markets. They are the primary generators of the knowledge that will result in success. As they travel along the path there are others who need to play a role of assessing and helping. Senior or experienced people need to look at the efforts of the commercialization team and help the team make the right decisions. These advisory people must be aimed in the same direction as the team. Their role is to advise and steer the effort along the path. They must be committed to success but still possess expert advice. This critical phase of the activity not only generates knowledge that will be used at future steps but also builds prototype products and processes. It should develop market knowledge by using test panels to determine what reception the product concept will have in the marketplace. Developing the *detailed commercial concept* is essential to going to the next step.

The next step is implementation. The team needs to be expanded to include people who can build the manufacturing process, procure the desired components, provide prototype product for test marketing, fine-tune the product, determine market strategies and channels, and perform the many other activities that need to be developed to introduce a new product into the marketplace. This *implementation* extends well into introduction of the product into the marketplace. The members of the team change as various tasks are complete, but a core group must exist. This core group is a combination of change agents and operational people. The initial offering is a time when problems will appear and need resolution. The change agents must solve these problems and further expand the knowledge base of this new supply chain.

With a supply chain established, activity shifts to the approach listed previously for change to defined operational supply chains. This activity needs to exist to make sure that commercial success is realized and that an agile response is made to unexpected changes in the marketplace. These unexpected changes will be more frequent in the early life of a product or product family. Indeed, it will be more difficult to anticipate the unexpected. Since marketplace success can exceed expectations, stretching the manufacturing capability and the supply chain may be required. If sales are not realized, then the way the product is placed in the market may need to be corrected. It may require adding functionality to the product. In an extreme case, it may require scaling down the supply chain consistent with the volume of product being sold. All the unexpected changes must be reacted to in the most agile way possible.

The *pathway to operational change* should be followed to take this new supply chain from the introductory level to the operational integrity required. Change in the supply chain is a continuing effort as progress is made and knowledge is developed. This is measured by improvement on the learning curve of the supply chain.

There are a few other bold changes that organizations, if properly climatized to change as a way of life, can react to. When the leadership of Motorola determined that having six sigma quality was a requirement for their survival, it meant that the whole organization needed a different focus. When an organization changes or improves its workflow processes, the leadership is committing all the people in the organization to change.

Summary of Pathways— Operational and New Technology

Pathways to change take on a different look depending on the state of knowledge in the supply chain. In ongoing activities change becomes a continuous improvement process. This process can be managed with the operational team. It should not be looked at as a quick fix and then forgotten. It should be viewed as a continuous journey with the team improving its ability to manage change as results are attained. After the team is comfortable with change, then more dramatic change can be incorporated into the process. The way the supply chain does business can be challenged and result in improvements. New commercial activities—such as new products, processes, or markets—can be incorporated. The pathways of change are the same. The approach to operational change is proven to work, and it is effective. I have personally used it and feel that the process is proven. It should be considered for every change program. The process is summarized in Figure 13.2.

Figure 13.2   Pathway to Operational Change: Continuous Improvement, Bold and Creative Change, and Corporate-Directed Change

1. Assess **Customer Focus and Responsiveness** as a determination of the need for change.
2. Define the **Supply Chain and Its Operational Team** as the focus of the improvement process.
3. Develop the **Improvement Process** that will be used to execute the change.
4. Develop an **Understanding of the Vision and Culture** of the firm and use it as a foundation for the change process.
5. Develop an action-oriented **Mission Statement** that can be used to set goals and define actions.
6. Translate the statement into an **Initial Change Program** using the supply chain team.
7. Involve expertise centers in the development of the **Change Program with the First Year Defined in Detail.**
8. **Implement the First-Year Program** involving all the change agents from both the operational team and the expertise centers.
9. Assess the **Results of the First Year** as a basis for the ongoing program.
10. Develop a new **Five-Year Program** with emphasis on the new first year.

> **Figure 13.2** *continued*
>
> 11. **Review Results and the New Program** with leadership to gain their support.
> 12. With two to three years of progress on the continuous improvement program, the leadership and others engage in **Entrepreneurial Brainstorming.**
> 13. Develop a **Bold and Creative Change Program** aimed at introduction of new products, processes, and markets into the firm. Expand to include new business practices.
> 14. Use the **Pathways of Change** concept to make the bold and creative change.

Change of continuous improvement leads to entrepreneurial brainstorming and significant levels of change in programs. This activity takes advantage of the change capability that has been developed by the supply chain team. The type of change at this stage can be more dramatic and thus make a more significant shift. This can involve bringing new business practices into the supply chain that were considered too bold in the past. It can put a completely new twist on the way the supply chain operates. It may be desirable to shift to a more increased customer responsiveness. Products could be tailored to the customer's demands and produced specifically for the defined application. Packaging could focus more on individual customer needs, and private labels could become a part of the offering. Emphasis would be on additional growth by providing a more agile or custom response to the customer. The Hewlett-Packard shift to the concept of postponement was driven by a corporate initiative that said the supply chains needed to be different to compete. Continuous improvement not only sets the stage for change but also becomes the pathway to make more dramatic change.

Historically the shift from mass production to lean could be viewed as learning to make change. The shift to agile or custom takes advantage of the changes involved in getting to lean. It takes the next step—that of making the supply chain more customer responsive and offering service or product that is custom tailored. Agility in delivering the custom-tailored service or product will make it cost competitive also.

Another change that needs to be considered is when an organization desires to develop a new supply chain with a new market, product family, and manufacturing and delivery capability. This type of change requires special consideration. The complex nature of this change requires a discipline that comes from the understanding of the magnitude of the task. Where a new product could be introduced effectively into the operational change process, this process will require the development of operational capability. The following process (see Figure 13.3) is aimed at making the effort disciplined and improving the chance of success.

Technology change and the development of a new commercial product line with supporting supply chain is not a low-risk task and must be done with discipline and care. The commitment of resources is

---

Figure 13.3   Pathway to Technology Change—Agile and Custom

1. The process begins with a **Market Assessment** either driven by a perceived need in the marketplace or by an internal technology.
2. With the market or application defined, the **Product Technology** can be determined with focus on both what is available and how far it is developed.
3. The product technology leads to a market, manufacturing process, and product **Detailed Commercial Concept.**
4. Plans and activities need to be put in place to address the commercial opportunity, and **Implementation** needs to proceed. Parallel activity needs to exist for developing the new product and the new process and for understanding the new market.
5. Once the product is in the marketplace, then **Reassessment** needs to occur and the pathways to operational change instituted with this new supply chain.

much greater than with the continuous improvement of an existing operation, so the benefits must exist to make this risk acceptable.

This chapter has defined two pathways to change: one for improvement in operational activity and another in bolder step change. Each has its place in moving the firm to a more agile and custom capability. The next chapter will summarize what has been presented. It will highlight the important concepts that make up the shift to agility and mass customization.

CHAPTER 14
Conclusion

History has shown that the industrial world is in a continuous state of improvement as it strives to become more effective and competitive. Progress has occurred with both big steps and with a lot of little ones. The big steps get recognition, but the many smaller contributions usually make the most progress. People like Whitney, Ford, Anderson, and Toyoda will be recognized for the many concepts that they pioneered. Others took those concepts and made them even more useful.

Today, mass production continues to be improved. Lean manufacture and the principles that define it are being implemented at many companies around the world. Lean production has gone beyond its origins in Japan to an essential part of doing business competitively worldwide. The focus on the machines needed to produce the product as well as on the product has impacted the electronic, automobile, and many other industries. A large number of opportunities still exist in Japan and many other parts of the world for continuing technology advances. Competitiveness will change as industry after industry incorporates lean manufacturing.

Another increase in competitiveness has occurred in recent years. Productivity improvement has focused on the blue collar production worker in the past. The focus is presently on the white collar worker who has been running the supply chain. The advance of integrated information systems has had a significant effect on these workers with productivity improving from 20 to 40 percent. This change in information systems has changed how business is done and has eliminated jobs. Products like SAP R-3 allow for data to be entered once and then used by all who need it. This eliminates the handling of data from one software to another. The software that is a part of this system is much faster and more robust, allowing for better decisions to be made. The net effect is a loss of jobs and an enhancement of productivity.

Another effect that has occurred recently is the outsourcing of work that the enterprise is not expert in. More focus has been on core competitiveness and less on doing everything. This shifts the work to a contract firm that manages it more effectively than the enterprise could manage it themselves. Jobs in the enterprise are lost and lower-paying jobs are usually created in the contract.

Finally, competitiveness has resulted in the reassessment of what business an enterprise would like to retain and which it would give up. Decisions have been made to give up business that has been a part of the enterprise for years and refocus the energy toward a new direction. Normally the business is sold, but in a few instances, it is abandoned to the competitors. The result is a loss of jobs within the enterprise.

These three activities result in the streamlining of the enterprise and the improvement of productivity. Jobs are lost, and in many cases people are displaced. White collar productivity has been the focus and it will continue as more use of the computer occurs.

The industrial world has moved from crafts workers to mass production to lean production and currently is in the initial phase of another step in global competitiveness. That step is mass customization—giving people a product that is made just for them or seems to be that way. This new step does not mean that any of the other steps have been displaced. Each continues and improvement occurs as we learn more about how to conduct business competitively using all the tools developed in all the steps that apply to a situation. Agility is one of those tools that has been used more as industrial improvement has been occurring.

Agility is defined as *the ability to respond with ease to the unexpected*. It means that the *unexpected* has been *anticipated* and the *capability has been built* so that the response can occur with ease. As industry moves from the crafts era, where significant agility existed, to mass production and then to lean manufacturing it has had to respond to the unexpected. This can occur either in making stepped improvements or in conducting business. Agility has been a key factor in what a firm does. The move to mass customization will require an even more agile capability to respond, a capability that has been developed for the circumstance in which business will be conducted.

With the change to lean manufacturing, companies have seen the opportunity and benefit in doing business differently. The world is shifting to an increased customer focus. The importance of the customer to the success of the enterprise has been discovered, and a new respect has been given to this individual. The view of business as an extended supply chain that is integrated from the customers to the suppliers is a key element of current thinking. The extended supply chain is the fundamental unit of operations and management of a business. It needs to be viewed as an integrated material, information, and cash flow process. It needs to be looked at as having no waste.

Waste takes the form of material and time efficiency. Either can be wasted. The focus of the supply chain needs to be on the customer, and having waste in the supply chain does not provide the customer with the right-priced product in the time desired. With waste, the supply chain is not streamlined and efficient, and a lot of things stand in the way of being effective. Some examples are inventory that is idle and not working or adding value, product that was not made right the first time, or machines that are not organized and managed properly so that continuous flow manufacturing results.

A supply chain or operational organization that has a lot of waste cannot be made agile easily. Because of all the things that are done wrong, it is not possible to respond to a customer easily. An example might be when a commitment for delivery in three days is made on a product that has not yet been produced, and the first try to produce it does not work. Then it is hard to be agile or deal with mass customization. Agility requires making the operational supply chains effective.

To make the operational supply chain more effective, it is necessary to look at how change is made in corporations. Very few firms have a culture that promotes change. Many have a strong resistance to change. This needs to be understood for any firm, and an approach must be developed to enhance change in the operational extended supply chains. This is best done by using a disciplined change management process. The process itself must be agile, however, so that directions can change with ease as new information is acquired. Discipline is required to keep focused on the overall vision or mission that has been selected. The change process should get as much attention as successfully executing today's business. It is not something that can be treated lightly if it is to be a success.

A key element of successfully managing change is to agree on a mission, vision, or direction established for the extended supply chain. Business practices need to be clarified so that the capability that is developed is consistent with these practices. Specific changes need to be defined and prioritized, and each change needs to be included in the plan in a time sequence. The change process needs to be occurring while the supply chain is operating, that is, serving customers and making money. Initially the amount of change cannot be large. It should be a step toward the vision of the extended supply chain. The customers, markets, channels, and suppliers all need to be included in the change plan thinking. The breadth of the extended supply chain needs to be the focus of the improvements in the capability relative to materials, information, and cash flows.

Agility needs to become a focus early in the change activity. A too rigid and inflexible capability will be hard to change in the future. This does not mean that the capability cannot be disciplined or even automated. It must be consistent with the vision that defines the way business along the supply chain is going to be done. Likewise the information and cash systems must be changed consistent with the vision.

A very important capability that will get more attention than any others will be the improvement in the capability to reliably manufacture the product and deliver it to the customer without an error. To move to lean or mass customization the capability must exist to operate as flawlessly as possible. Departure from reliable operation causes things to happen that detract from being as effective as possible. Effectiveness leads to competitiveness.

CONCLUSION ■ 251

Making the operational extended supply chain as effective as possible is like building the foundation for a building. Many things can happen with a sound footing. The further, or bolder, change in the way business is done may be possible with the improved capability of the supply chain. A shift of part of the business to mass customization might prepare the way, by building capability, for using this technique in a broader fashion across the product line. Driving the lead time down in the supply chain, with the right building of capability, can enable more agile response to customers, such as a make-to-order. This requires a different approach than make-to-inventory.

Another part of the organization will benefit from the improved extended supply chain. It also benefits from the shift to a change-oriented culture. This part of the organization is the one that builds new capability in the product, market, and process for the enterprise by looking at existing supply chains and the technology that supports them as the foundation to add a new capability to the organization. New technology is the lifeblood of an organization, and it is essential that it be implemented in as short a time as possible. It is also important that it succeeds and meets expectation relative to the benefits it will bring to the firm. There are three dimensions of this type of commercialization, which are depicted on the Three Dimensions of Commercialization cube illustrated in Chapter 8 of the text.

New products represent one of the dimensions of commercialization and must be designed to meet the customer's expectation but must also be designed to be produced or manufactured. Another dimension is new processes; they must be developed to produce the product or intermediate that fits the existing applications. Taking a product into a new market is another of the dimensions. It can be either within the existing business region but with new applications or outside the region in either new or existing applications. Each will require special thinking and an organized approach to the new customers and/or the new application. Going global is a part of this approach and that is a more difficult challenge.

The challenges increase when more than one of the dimensions is involved with the change. A coordinated effort is required among the people involved with the three dimensions. Each group has a reason not to trust the others. Multidimensional commercialization projects work best if all the people are assigned to the commercialization team and have their loyalty with this effort. The biggest challenge seems to exist when the three dimensions are all new. New product, new process, and new market with lots of new technology are projects that succeed if they are positioned high in the corporation where the proper risks can be taken and the right actions decided upon. The ability to be successful in product, process, and market commercialization benefits from the success of the firm in operational supply chain improvement. The dimensions come together and represent the overall change activity in the enterprise. The people in the corporation have adapted to change and they have made it a part of the culture. This enables it to be successful as long as there is support for the change that is being asked for.

In any business or undertaking it is the people who are the key to success, who make the difference. To be successful, the relationships of people must be understood. These relationships can vary from those that are very agile and adaptable to ones that are very rigid or, worse, apathetic. Industrial activity needs to have relationships that allow people to cooperate to enhance a firm's competitiveness. The actions of the leadership must promote cooperation. This requires "playing the game fairly" and treating all individuals with respect. Cooperation is essential for operation of a supply chain. It is also essential for the improvement of an operational supply chain as well as in commercialization of new processes, new products, and new markets. Commercialization cannot exist in an effective fashion if the people are not all cooperating to make the venture successful.

In agile or custom manufacturing, people must be responsive to unexpected changes. They must build capability that is required. People are what make operational extended supply chains better. They are what make commercialization successful. They provide the change and learning necessary to make lean or mass manufacturing operate and improve. People are knowledge workers and must be treated in this fashion to make them successful. Because they make the difference, people must be nurtured and developed, respected and treated fairly. They have the knowledge that makes the firm what it is and must be treated as an asset, an equivalent asset to the capital assets that provide cash and physical facilities.

Agility is a tool that applies to all the ways that business is done. It has characteristics that make it important to competitiveness. These characteristics are hard to see or build when a supply chain is not operating very effectively. Initial change needs to be directed toward making the supply chain more effective. This can be followed by bolder change that makes it more agile and possibly even takes the supply chain to mass customization. The improvement of a supply chain helps to build a change in culture that sets the environment for bolder improvements. New capability brings agile processes into the supply chain. A broader product technology allows for custom formulation to fit the application needs in a make-to-order fashion. The pathways to agility represent the avenues for change to a more competitive company. They focus on providing the atmosphere for successful change and the development of a change culture for the corporation. They stress the improvement of the supply chain toward a more lean approach to operation. They stress the building of an integrated supply chain with reengineered workflows and computer systems that make this happen with ease. They stress that agility is a key tool for the competitive enterprise.

Pathways to Agility: Mass Customization in Action has provided the reader with both the processes needed to make change effectively and the tactics associated with improvement. The concepts and the directions presented are from proven examples or personal experiences. The changes discussed will help an enterprise or a supply chain attain a higher level of efficiency, effectiveness, and competitiveness. All readers are encouraged to make the effort to make the difference.

References and Suggested Reading

Camp, Robert C. *Business Process Benchmarking—Finding and Implementing Best Practices*. Milwaukee, WI: ASQC Quality Press, 1995.

DeMeyer, Arnoud, Hiroshi Katayma, and Jat S. Kim. *Building Customer Partnerships as a Competitive Weapon: The Right Choice for Globalizing Competition?* Manufacturing Round Table—Report on the 1996 Global Manufacturing Futures Survey (Japan, Europe and the United States), Research Report Series, Boston University, October 1996.

Dove, Rick. *An Ear for Strategy*. Automotive Manufacturing & Production, January 1997.

Feitzinger, Edward, and Hau L. Lee. Mass Customization at Hewlett-Packard: The Power of Postponement. *Harvard Business Review*, January–February 1997.

Gilmore, James H., and Joseph B. Pine II. "The Four Faces of Mass Customization." *Harvard Business Review*, January–February 1997.

Goldman, Steven L., and Kenneth Preiss, ed.; Roger N. Nagel, and Rick Dove, Principal investigators, with 15 industry executives. *21st Century Manufacturing Enterprise Strategy: An Industry-Led View*. 2 volumes. Iacocca Institute at Lehigh University, Bethlehem PA, 1991.

Goldman, Steven L., Roger N. Nagel, and Kenneth Preiss. *Agile Competitors and Virtual Organizations: Strategies for Enriching the Customer*. New York: Van Nostrand Reinhold, 1995.

Grenier, Ray, and George Metes. *Going Virtual—Moving Your Organization into the 21st Century*. Upper Saddle River, NJ: Prentice-Hall, 1995.

King, Julia. "Sharing IS Secrets—Retail Projects Cut Supply Chain Costs." *Computer World*, 23 September 1996.
Martin, Justin. "Are You as Good as You Think You Are." *Fortune Magazine*, 30 September 1996.
Morrison, Ian and Greg Schmid. *Future Tense: The Business Realities of the Next Ten Years*. New York: William Morrow, 1994.
Moskal, Brian S. "Son of Agility." *Industrial Week*, May 15, 1995. Cleveland, OH: Penton Publishing.
Patterson, Rusty. "Firms Need to Be Agile," Commentary. *The Philadelphia Inquirer*.
Pisano, Gary P., and Steven C. Wheelwright. "The New Logic—High Tech R&D." *Harvard Business Review*, September–October 1995.
Porter, Michael E. "What Is Strategy?" *Harvard Business Review*, November–December 1996.
Rayport, Jeffrey F., and John J. Sviokla. "Exploiting the Virtual Value Chain." *Harvard Business Review*, November–December 1995.
Sellers, Patricia. "The Dumbest Marketing Ploy." *Fortune Magazine*, 5 October 1992.
Upton, David M. "What Really Makes Factories Flexible?" *Harvard Business Review*, July–August 1995.
Verity, John W. "Information Management—Clearing the Cobwebs from the Stockroom." *Business Week*, October 21, 1996.
Wagner, Betsy. "Computers—Another Pinnacle." *U.S. News & World Report*, 30 December 1996/6 January 1997.
White, Joseph P. "Next Big Thing—Re-Engineering Guru Take Steps to Remodel Their Stalled Vehicles." *Wall Street Journal*, 26 November 1996.
Womack, James P., and Daniel T. Jones. *Lean Thinking—Banish Waste and Create Wealth in Your Corporation*. New York: Simon and Schuster, 1996.
Womack, James P., Daniel T. Jones, and Daniel Roos. *The Machine That Changed the World: The Story of Lean Production*. New York: Rawson Associates, 1990.

Index

A
A&V Conferencing
   core competencies of, 209–210
   key factors in relationships, 210
   outsourcing at, 208–209
Action-oriented mission statement, 233
Action strategies, 41
Agile automation
   of cash flow, 128–129
   discipline in, 129–131
   of information and decision-making processes, 125–127
   of material handling, 124–125
   overview of, 119–121
   of processes or machines, 121–124
Agile manufacturing, 21–22
   examples of, 32–37
Agile Manufacturing Enterprise Forum. *See* Agility Forum
Agile relationships
   cooperation in, 200–202, 204–211
   employee-oriented, 218–222
   external, 211–215
   internal, 215–218
   overview of, 199–200
   summary of, 222–223
   in supply chains, 203–204
Agility
   corporate culture for, 20–22
   defined, xv–xvi, 248
   in discrete industry, xvi
   history of, 3–4
      lean production, 6–9
      mass production, 4–6
   in process industry, xvi–xviii
   role in supply chain, 29–32 (*see also* Supply chains)
   U.S. government's role in development of, 9–16
   See also Agility pathways; Capability
Agility Forum, 13, 15–16
Agility pathways
   establishing direction of change, 225–226
   to new technology change, 236–242
   to operational change, 226–236
   overview of, 224–225
   summary of, 242–246

Agriculture, agility in, 3
AMTEX, 10, 11, 205
Anderson Window, 32–33, 36–37
Artificial demand, 103–104
Artzt, Edwin, 137
Asia, market expansion into, 189, 190
Asset utilization, 110
    through agile automation, 128
Automation, 118–119. *See also* Agile automation
Automobile industry
    custom manufacturing in, 33–34
    extended supply chain in, 53–56
    lean production in, 7–8
    mass production in, 4–6
    outsourcing in, 78–79
    product introduction, 178
    USCAR program for, 12, 205
    See also Ford Motor Company; General Motors; Porsche; Toyota

B
Bar coding, 127
Barker, Joel, 24–25, 100
Benchmarking, 15
Boeing
    agile relationships of, 201–202
    agility and customization at, 34–35
    key factors in relationships, 210
Bossidy, Larry, 50–51
Brainstorming, entrepreneurial, 235
Breakthrough thinking, 227
Budgets, development of, 41
Building Customer Partnerships as Competitive Weapon (DeMeyer, Katayama, and Kim), 47
Business organization, structure of, 49–51
Business practices, improvement of, 26–27
Business relationships
    external, 211–215
    internal, 215–218
    See also Agile relationships
Business supply chain, 51–57. *See also* Extended supply chains; Supply chains

C
Camera industry, product introduction, 179. *See also* Kodak
Candy industry, 115–116
Capability, xvi
    assessment of, 100–104
    cash flow, 113–114
    and customer orders, 104–110
    information flow, 112–113
    material flow, 111, 112
    of new markets, 193–195
    of new processes, 163–166
    of new products, 180–181
    overview of, 99–100
    product and service quality, 114–116
    summary of, 116–117
Cash flow, agile automation of, 128–129
Cash flow capability, 113–114
Catalog sales, 67, 170
Central America, market expansion into, 190–191
Change
    establishing direction of, 225–226
    necessities for, 24–32
    random prioritization of, 22–24
    See also Operational change; Technology change
Change management, 16–18
    in corporate culture, 20–22
Change programs, 233–234
Channels. *See* Distribution channels
Charter, development of, 38
Chemical industry
    Computational Fluid Dynamics (CFD), 95
    importance of people in, 221–222
    make-to-order supply chain, 108–109
    PI system, 89–90
    See also Dow Corning; Midwest Chemical
China
    developing business relationships in, 211–212
    market expansion into, 190

INDEX ■ 259

Chinese Overseas Network, 190
Commercialization
   dimensions of, 136–146
   mass customization as part of, 146–150
   overview of, 135
   process dimension of, 140–141, 153–157
   process for new products, 173–178
   product dimension of, 137–139, 169–170
   speed to market, 178–180
   summary of, 150–151
   See also New markets; New processes; New products
Competitors, cooperation between, 204–211
Computational Fluid Dynamics (CFD), 95
Computer-aided design (CAD), 88
Concurrent engineering, 8
Continuous improvement, 226–227
   overview of, 232–235, 243–244
   at Specialty Resin and Lubricants, 227–232
Contract relationships, 214
Controlling/cost accounting module, 93–94
Cooperation
   in agile relationships, 200–202
   between competitors, 204–211
Cooperative Research Act (1984), 10, 12, 162, 204
Core competencies, 77–78
   of A&V Conferencing, 209–210
   of Boeing, 201–202
Corporate culture, 20–22
Corporate-directed change, 243
Crafts workers, agility of, 4
Culture, 217–218, 219
   and continuous improvement, 233
   corporate, 20–22
Customer focus and responsiveness, 232
Customer orders, 104–105
   agile automation of, 126–127
   customization of, 105–107
   satisfying, 107–110

Customers
   responsiveness to, 73
   use patterns of, 103
   See also Extended supply chains; Supply chains
Customer service, 110
Customization, 105–107. *See also* Mass customization
Customized products, 170–173
Custom manufacturing, examples of, 32–37. *See also* Mass customization

D

Data systems. *See* Integrated information systems
Decision-making processes, 125–127
Delphi, custom manufacturing at, 34
Demand, artificial, 103–104
Detailed commercial concept, 241
Dimensions of Commercialization cube, 136–137. *See also* Commercialization
Discipline, in automated agility, 129–131
Discount stores, 66–67
Discrete industry, agility in, xvi
Distribution channels, 67–68. *See also* Extended supply chains
Distributors, role of, 76
Dove, Rick, xvi, 13
Dow Corning, xvii
   implementing change at, 23–24
   market expansion into Japan, 190
   new commercialization at, 143, 192
"Dumbest Marketing Ploy, The" (*Fortune*), 103, 137–139
Du Pont, 11

E

Eastern Europe, market expansion into, 191
Electronic commerce, 68
   and cash flow, 113–114
Electronics industry, agility in, 10–11
Employees, agile relationships with, 218–222

England. *See* United Kingdom
Enterprise resource planning (ERP) software, 91. *See also* Integrated information systems
Entrepreneurial brainstorming, 235
Europe, market expansion into, 191
Extended supply chains, 52, 66–69
  customers, markets, and channels, 69–76
  role of suppliers in, 77–79
  summary of, 80–82
  in U.S. auto industry, 53–56
Eye Tailor, 105

F
Farmers, agility of, 3
First-year program, 234
Five-year program, 233–234
  at Specialty Resin and Lubricants, 230
Ford, Henry, 4–5, 219
Ford Motor Company, 7
  concurrent engineering at, 8
  outsourcing at, 79
Foresight, 17–18
Future Tense (Morrison and Schmidt), 17

G
Gap analysis, 227
Garment industry, new products in, 170–173
General Motors
  new process at, 155–156
  outsourcing at, 79
Geraci & Associates, Inc., 38. *See also* Strategic mapping
Germany, developing business relationships in, 212
Global market expansion, growth through, 189–191
Goals, transforming mission statement into, 41
Going Virtual (Grenier and Metes), 237
Goldman, Steven, 13
Government. *See* U.S. government

Great Britain. *See* United Kingdom
Green Book (Dow Corning), xvii, 23
Gretzky Effect, 17–18

H
Hall, Alphonso, 155
Hertz, 106
Hewlett-Packard (HP)
  key factors in relationships, 210
  mass customization at, 147, 148–149, 206–207
Hunter-gatherers, agility of, 3

I
Imai, Maasaki, 164
Implementation, 241
Improvement activity, 233
Industrialization, 19–20. *See also* Mass Production
Information flow capability, 112–113
Information systems
  agile automation of, 125–127
  integrated with workflow processes, 216
  in supply chains, 63, 64
  See also Integrated information systems
Initial change programs, 233
Integrated enterprise, 91–94
Integrated information systems, 68
  integrated enterprise, 91–94
  IPPD, 87–89
  manufacturing process information, 84–86
  overview of, 83–84
  process control, 89–91
  scientific computing, 95–97
  of Specialty Resin and Lubricants, 231
  summary of, 97–98
Integrated operational system, modules for, 92–94
Integrated Product and Process Data System (IPPD), 87–89
Integrated Supply Chain Management (ISCM), 47–48

INDEX ■ 261

Inventory
    in make-to-stock system, 70–71
    reduction of, 73
    in supply chains, 58, 60
Iwata, Yoshiki, 164

J
Jackson and Perkins, 105, 169
Japan
    auto industry (*see* Automobile industry)
    market expansion into, 189, 190
    Ministry of Industry and Trade (MITI), 205
JD Edwards, 68

K
Kaikaku, 165
Kanban, 165
Kim, Jay S., 48
Knowledge points, 237, 238
Kodak
    new product development process, 177–178
    product introduction, 179

L
Land's End, 104–105
Lean production, 6–9
Lean Thinking (Womack and Jones), 9
Lehigh University, 13
Levi Strauss, customized products at, 170–173

M
Machines, agile automation of, 121–124
Machine That Changed the World, The (Womack, Jones, and Roos), 9
Make-to-inventory system, 70–71
Make-to-order system, 71–72, 101, 105
    in chemical industry, 108–109
    effect on supply chain, 109–110
Manufacturing. *See* Agile manufacturing; Custom manufacturing; Lean production; Mass customization; Mass production

Manufacturing process information, 84–86
Manufacturing technology, 7
Market assessment, 240
Markets
    global expansion of, 189–191
    regional expansion of, 188
    See also New markets
Mass customization, xv, 206
    at Levi Strauss, 171–172
    as part of commercialization, 146–150
    See also Custom manufacturing
Mass production, 4–6
Material flow capability, 111, 112
Material handling, agile automation of, 124–125
Measurement system, development of, 28
Megatrends Asia (Naisbitt), 190
Mexico, market expansion into, 190–191
Midwest Chemical
    key factors in relationships, 210
    outsourcing at, 208
Mind-set. *See* Culture
Mission statement
    action-oriented, 233
    development of, 39–40
    of Specialty Resin and Lubricants, 228–229
Modules, 92–94
    for mass customization, 147–148
Morrison, Ian, 17
Moskal, Brian, 155
Motorola
    agile automation at, 124–125
    agile manufacturing at, 33
    change management at, 20–22
    new commercialization at, 143, 192
    new process technology at, 160

N
Nagel, Roger, 13
Naisbitt, John, 190
Nakao, Chihiro, 164, 165

New capability. *See* Commercialization
New market capability, 193–195
New markets, 139–140
  change process for, 236–242
  global expansion, 189–191
  new application in, 183–187
  new product with new process in, 141–146, 191–193
  overview of, 182–183
  regional expansion, 188
New process capability, 163–166
New processes, 140–141
  change process for, 236–242
  as dimension of commercialization, 153–157
  new products in new market with, 141–146, 191–193
  overview of, 152
New process technology, 158–163
  summary of, 166–167
New product capability, 180–181
New products, 137–139
  change process for, 236–242
  development of, 173–178
  as dimension of commercialization, 169–170
  with new process in new market, 141–146, 191–193
  overview of, 168–169
  speed to market, 178–180

O
Ohno, Taiichi, 5
Operational change
  pathway to, 226–236
  summary of, 242–245
Oracle, 68
Organizational culture. *See* Culture
Outsourcing, 206
  at A&V Conferencing, 208–209
  in automobile industry, 78–79
  at Midwest Chemical, 208

P
Partnerships, 214
Pathways. *See* Agility pathways

PeopleSoft, 68
Performance measurement. *See* Measurement system
Pick-and-ship process, 104–105
Porsche, 163–165
Porsche, Ferry, 163
Porshe Verbesserungs Process, 164
Preiss, Kenneth, 13
Process control, 89–91
Processes
  agile automation of, 121–124
  workflow, 216
  See also New processes
Process industry, agility in, xvi–xviii
Process Information (PI) system, 89–90
Process technology, new, 158–163
Proctor & Gamble (P&G), 103, 137–139
Production planning module, 93, 94
Product life cycle, 169. *See also* New products
Product quality, 114–116
Products, customized, 170–173. *See also* New products
Product technology determination, 240

Q
Quality management module, 92–93

R
Regional market expansion, growth through, 188
Relationships. *See* Agile relationships
Retail channels, shifts in, 66–67

S
SAP, 68, 91, 92, 94
Schmidt, Greg, 17
Scientific computing, 95–97
Search for a New Manufacturing Paradigm (Kim), 48
Sematech, 10, 79, 162, 204–205
Service quality, 114–116
Shingijutsu group, 164
Software. *See* Integrated information systems

INDEX ■ 263

Solectron, agile automation at, 123–125, 126, 127
Sony, 179
South Korea, market expansion into, 189, 190
Specialty Resin and Lubricants (SRL), continuous improvement at, 227–232
Specialty stores, 67
Strategic mapping, 37–38
    action strategies, 41
    budgeting and assessing, 41–43
    charter development, 38
    goals, 41
    mission statement, 39–40
Strauss, Levi, 170–171
Suppliers, 77–79. *See also* Supply chains
Supply chain capability. *See* Capability
Supply chains
    agile relationships in, 203–204
    changes in, xv
    continuous improvement of, 232–233, 234
    improvement of, 27
    in mass customization approach, 148
    measurement system for, 28
    overview of, 47–49
    role of agility in, 29–32
    time factors in, 57–64
    See also Business supply chain; Extended supply chains

T

Taiwan, market expansion into, 189, 190
Technik, 163
Technology. *See* New process technology

Technology change
    pathway to, 236–242
    summary of, 245–246
Textile industry, AMTEX program for, 10–12, 205
Time, in supply chains, 57–64
Toyoda, Eiji, 5
Toyota Motor Company, 5
    outsourcing at, 78, 79
Toyota Production System, 5–6
Tracking systems, 127
21st Century Manufacturing Enterprise Strategy (Goldman and Preiss, eds.), 13, 14

U

United Kingdom, developing business relationships in, 212
United States
    auto industry (*see* Automobile industry)
    developing business relationships in, 212–213
    development of agility in, 9–16 (*see also* AMTEX; Cooperative Research Act; USCAR)
    industrial competitiveness in, 9–13
USCAR, 10, 12, 205
Use patterns, 103

V

Value-added time, 57–60
Vertical integration, 69
Victoria Secret, 170

W

Whitney, Eli, 4
Wiedeking, Wendelin, 164, 165
Workflow processes, 216